Why Do Fractals Matter

John Archibald Wheeler (b. 1911), protégé of the quantum pioneer Niels Bohr and friend of Albert Einstein, has been at the cutting edge of 20th-century physics, cosmology and quantum theory. **Ian Stewart** is a respected Professor of Mathematics at Warwick University. They are among the many scientists agreed that fractal geometry is a revolutionary breakthrough in our comprehension of reality.

3

A Smooth World or a Rough One?

Plato sought to explain nature with five regular solid forms. Newton and Kepler bent Plato's circle into an ellipse. Modern science analysed Plato's shapes into particles and waves, and generalised the curves of Newton and Kepler to relative probabilities – still without a single "rough edge". Now, more than two thousand years after Plato, nearly three hundred years after Newton, Benoît Mandelbrot has established a discovery that ranks with the laws of regular motion.

Professor Eugene Stanley, Center for Polymer Studies, Department of Physics, Boston University

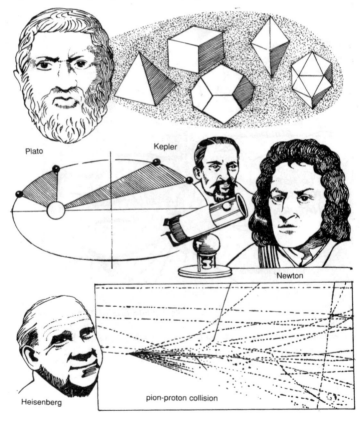

Plato

Kepler

Newton

Heisenberg

pion-proton collision

Benoît Mandelbrot

The world that we live in is not naturally smooth-edged. The real world has been fashioned with rough edges. Smooth surfaces are the exception in nature. And yet, we have accepted a geometry that only describes shapes rarely – if ever – found in the real world. The geometry of Euclid describes ideal shapes – the sphere, the circle, the cube, the square. Now these shapes do occur in our lives, but they are mostly man-made and not nature-made.

5

The Texture of Reality

Nature deals in non-uniform shapes and rough edges. Take the human form. There is a certain symmetry about it, but it is, and has always been, indescribable in terms of Euclidean geometry. It is not a uniform shape. This is the issue. What has been missing from the scientific repertoire until very recently has been a way of describing the shapes and objects of the real world.

The word "fractal" was coined in 1975 by the Polish/French/American mathematician, **Benoît Mandelbrot** (b. 1924), to describe shapes which are detailed at all scales. He took the word from the Latin root **fractus**, suggesting fragmented, broken and discontinuous.

Fractal geometry is the geometry of the irregular shapes we find in nature, and in general fractals are characterized by infinite detail, infinite length, and the absence of smoothness or **derivative**.

The Origins of Fractals

Fractal geometry is an extension of classical geometry. It does not replace classical geometry, but enriches and deepens its powers. Using computers, fractal geometry can make precise models of physical structures – from sea-shells to galaxies.

We will now trace the historical development of this mathematical discipline and explore its descriptive powers in the natural world, then look at the applications in science and technology and at the implications of the discovery.

Classical Geometry

Euclid of Alexandria (c. 300 BC) laid down the rules which were to define the subject of geometry for millennia to come. The shapes that Euclid studied – straight lines and circles – proved so successful in explaining the universe that scientists became blind to their limitations, denouncing patterns that did not fit in Euclid's scheme as "counter-intuitive" and even "pathological".

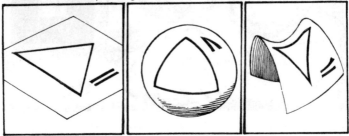

A steady undercurrent of ideas, starting in the 19th century with discoveries by **Karl Weierstrass** (1815–97), **Georg Cantor** (1845–1918) and **Henri Poincaré** (1845–1912), led inexorably towards the creation of a whole new kind of geometry, with the power to describe aspects of the world inexpressible in the basic language of Euclid.

The Calculus

Johannes Kepler (1571–1630) was the first to realize that planets followed elliptical orbits, not perfect circles. **Edmond Halley** (1656–1742) guessed that elliptical orbits could be explained, by analogy with light, using an inverse square law.

Sir Isaac Newton (1642–1727) derived a new method of reasoning based on the idea of vanishingly small quantities, or **infinitesimals**, in order to tame the complex motions of projectiles and planets and arrive at his celebrated theory of universal gravitation. The calculus was conceived simultaneously by Newton and **Gottfried Wilhelm Leibniz** (1646–1716). Leibniz developed the clearest formulation of the calculus, including the notation which is used to this day.

The twin tools of the calculus are **differentiation** and **integration**. Differentiation gives the derivative, or rate of change, of a variable. Rate of change is the key. For example, inflation is the rate of change of prices: the first derivative of average prices. Velocity is the rate of change of position over time: the first derivative of position. The second derivative of position, the rate of change of velocity, is called acceleration.

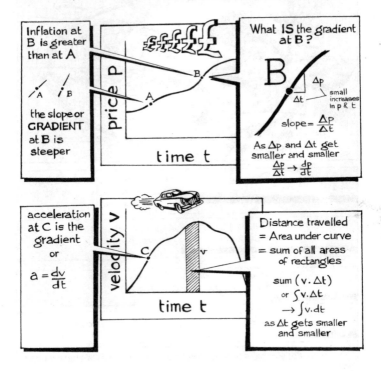

Inflation at B is greater than at A

the slope or GRADIENT at B is steeper

What IS the gradient at B?

slope $= \frac{\Delta p}{\Delta t}$

small increases in p & t

As Δp and Δt get smaller and smaller

$$\frac{\Delta p}{\Delta t} \rightarrow \frac{dp}{dt}$$

price P — time t

acceleration at C is the gradient or

$$a = \frac{dv}{dt}$$

velocity v — time t

Distance travelled
= Area under curve
= sum of all areas of rectangles

sum $(v . \Delta t)$

or $\int v . \Delta t$

$\rightarrow \int v . dt$

as Δt gets smaller and smaller

Integration is the reverse operation. Future values of a variable can be found by integrating, or summing, its rate of change at each instant in time. Systems controlled by physical forces like gravity can be analysed in terms of their rate of change. Combining these changes defines the evolution of the system.

11

The Paradox of Infinitesimals

Newton's theory of infinitesimals is riddled with paradoxes. The question of the infinite divisibility of space had perplexed philosophers for thousands of years.

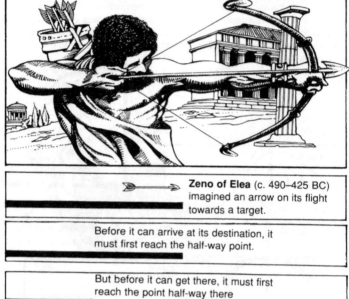

Zeno of Elea (c. 490–425 BC) imagined an arrow on its flight towards a target.

Before it can arrive at its destination, it must first reach the half-way point.

But before it can get there, it must first reach the point half-way there

. . . and so on.   This line of reasoning *ad infinitum*

apparently implies that the arrow can never leave its starting point.

Zeno had deduced the paradoxical nature of motion.

It took the efforts of **Augustin Louis Cauchy** (1789–1857) and his pupil Karl Weierstrass to banish infinitesimals. Until then, it was quite possible that the entire edifice of applied mathematics was founded upon a contradiction.

Effects of Calculus

Despite lacking a solid theoretical justification, the calculus was extremely successful. Newton's three laws of motion and the electromagnetic equations of **James Clerk Maxwell** (1831–79) flowed from this empowering discovery. The physical sciences were transformed. It was assumed that all phenomena could be understood in terms of these new techniques. **Pierre Simon Laplace** (1749–1827) claimed that given the position of every particle in the universe and its rate of change, we could predict the entire future of the universe forever in its every detail.

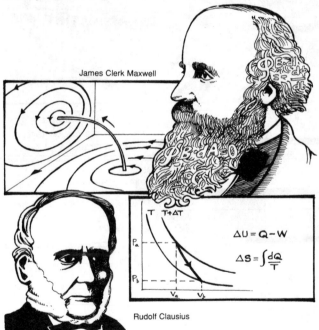

James Clerk Maxwell

$$\oint E \cdot dl = -\frac{d\Phi}{dt}$$

$$\oint B \cdot dA = 0$$

$$\Delta U = Q - W$$

$$\Delta S = \int \frac{dQ}{T}$$

Rudolf Clausius

The methods of the calculus apply wherever a curve is **smooth**. It was thought that any curve with bends or kinks could be split into separate smooth curves that would then succumb to calculus. That any curve could have only isolated corner points was not even questioned.

13

The First Fractal

The first mathematical fractal was discovered in 1861. Karl Weierstrass delighted in finding flaws in the arguments of others. His quest for absolute rigour led to his discovery of a nowhere differentiable continuous function: that is, a curve consisting *completely of corners*. It was just not possible to define its rate of change at any point. There was no smoothness anywhere. This was a shock to the scientists of the day.

A MATHEMATICIAN WHO IS NOT ALSO SOMETHING OF A POET WILL NEVER BE A PERFECT MATHEMATICIAN

Weierstrass's function was thought to be an aberration, a "pathological" product of the human mind resembling nothing to be found in nature. Weierstrass and Cauchy developed a new branch of mathematics called **Analysis**. Analysis was an attempt to bring a new rigour to mathematics. Precise notions of number and continuity were sought.

Explaining Numbers

If we start with the natural numbers: 1, 2, 3 etc. . . .

then add zero and the negative whole numbers like -5, we get the integers . . .

then add all numbers which can be written as fractions . . .

and numbers like √2 and π which can't be written as fractions . . .

and finally add numbers which include combinations of √-1, we end with the broadest set, complex numbers.

Firm Foundations and Sets

Mathematicians have always sought to put their discipline on a firm foundation. Analysis showed how all mathematics could be explained in terms of simple whole numbers. Could even the numbers be reduced to pure logic? There were several attempts, based on the idea of **sets**.

All mathematics can be defined in terms of sets. For example, numbers can be seen simply as properties of sets. Three is the common feature shared by all sets with three members. The circularity is removed by defining one set with three members, and appealing to the idea of one-to-one correspondence.

What Are Sets?

A set is a collection of things which can be thought of as a single object. This definition, by the way, excludes self-contradictory ideas like the "set of all sets"! Sets can contain other sets but they cannot contain themselves, for this way lies madness. **Bertrand Russell** (1872–1970), in his famous paradox, exposed the dangers of allowing sets to contain themselves.

DOES THE SET OF ALL SETS WHICH DON'T CONTAIN THEMSELVES CONTAIN ITSELF?

Interestingly, "set" has more distinct meanings than any other word in the English language. The *Oxford English Dictionary* lists 126 different definitions, many concerning groups or collections of objects. ("Mandelbrot" has only one meaning in any language, namely "almond bread".)

Cantor and the Continuum

A major stumbling block was the infinite. It required a leap of faith that many mathematicians were unwilling to take.

Georg Cantor, one of the pioneers of modern set theory, started out on a problem which continued to vex him for the rest of his life: the nature of the **continuum**. The continuum is the ideal infinitely divisible space conceptually required for a theory of continuous change.

Cantor's Diagonal Slash
In the left-hand column are the integers 1, 2, 3 etc. Matched to each is an arbitrary real number, which lies between 0 and 1.

I WILL SHOW THAT THE POINTS OF THE CONTINUUM ARE MORE NUMEROUS THAN THE NATURAL NUMBERS

Are there more real numbers than integers? Whichever real number appears on the right-hand side, can't we just find an integer to set opposite it, since there are infinitely many integers?

Well, look at the number on the diagonal.

It is 0.1704826 . . .

Now change this by subtracting one from each of the digits.
It becomes 0.0693715 . . .
This number will never appear on the list because it differs in at least
one of the decimal places from anything which appears there.
So, there are more reals than integers.

Cantor's argument implied the existence of *different types of infinity*. He
went on to develop a whole new theory of transfinite arithmetic,
convinced that he had uncovered a powerful new principle of reality with
profound physical and spiritual implications.

THERE IS NO ONE-TO-ONE CORRESPONDENCE BETWEEN THE REAL NUMBERS & THE NATURAL NUMBERS

THE NUMBER OF REAL NUMBERS IS ACTUALLY GREATER THAN THE NUMBER OF RATIONAL NUMBERS

Roger Penrose: Rouse Ball Professor of Mathematics, Oxford University

... AND IT'S NOT COUNTABLE

The Cantor Set

Cantor's quest for the meaning of continuity led him, in 1883, to the set that is now named after him, one of the first fractals to be studied mathematically. It was actually discovered by Henry Smith, a professor of geometry at Oxford, in 1875.

Take a line, remove the middle third, leaving two equal lines. Likewise remove the middle thirds from each of these two lines. Repeat this process an infinite number of times, and you are left with the Cantor set.

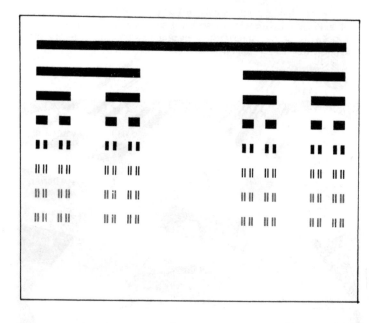

The Cantor set has no length or interior. In technical parlance, it has "zero measure". A randomly thrown dart is infinitely unlikely to hit it. It is "nowhere dense". Every part of it consists almost completely of holes.

And yet despite being nothing but totally disconnected points, it is *uncountable*. In fact, it contains as many points as the whole line it is carved from. Every point is an "accumulation" or "limit" point, meaning there are infinitely many other points from the set in any neighbourhood of it, no matter how small. Conversely, the Cantor set contains all of its limit points.

Any point arbitrarily close to the set must actually belong to it.

It is what Cantor called a "perfect" set. It is equal to its set of limit points. The existence of this infinitely divisible yet totally discontinuous set forced Cantor to refine his notion of continuity.

Peano's Space-Filling Curve

Mathematicians were searching for a definition of dimension. A bombshell burst around 1890 when **Giuseppe Peano** (1858–1932) discovered what was called a "space-filling curve". Peano had constructed an idealized curve which twisted in such a complex way that it visited every point in the entire plane.

IF YOU TRY TO PLOT IT..

.. IT WILL ACTUALLY FILL THE WHOLE PLANE OF THE PAPER

There was no point on the plane that Peano's curving line would not include.

This created an unpleasant situation for mathematicians: a bit like biologists being unable to define life, or philosophers consciousness. The very two-dimensionality of the plane lay in its set of points.

Was there a mapping that was both continuous and one-to-one? If so, the concept of dimension would have no **topological** meaning whatsoever. **Topology** is a mathematical study of geometrical properties and spatial relations that remain unaffected by continuous stretching, bending, twisting, etc.

Topological and Fractal Dimension

In 1911, **Luitzen Brouwer** (1881–1966) proved that there could be no such mapping. Dimension is a topological invariant. It cannot be altered by continuous deformation.

This gives a definition of the dimension of a shape or space, called the topological dimension. Another line of reasoning by **Felix Hausdorff** (1869–1942) led to a different take on dimension. Focusing on the manner in which shapes fill the space around them, Hausdorff derived a measure which extends our intuitive ideas of dimensionality. For more complicated shapes, other than normal Euclidean objects, Hausdorff's approach gave a fractional dimension. This allows for the intriguing possibility of **one-and-a-half-dimensional** objects.

How can a shape lie in-between dimensions? By being **fractal**.

Felix Hausdorff

Self-Similarity

Visually, it is apparent that the Cantor set consists simply of two small copies of itself. This property is known as **self-similarity**.

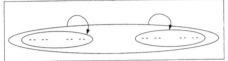

Self-Similarity of the Cantor Set

The Cantor set is a collection of two exact copies of the entire Cantor set scaled down by the factor 1/3.

Any part of a straight line is itself a straight line, identical to the whole line except for a scaling factor. Most Euclidean shapes do not share this property.

AN ARC OF A CIRCLE IS NOT ITSELF A CIRCLE

A SIDE OF A TRIANGLE IS NOT TRIANGULAR

YET IN NATURE SUCH SELF-SIMILARITY ABOUNDS

TREES, CLOUDS AND MOUNTAINS ALL RESEMBLE SMALLER PARTS OF THEMSELVES

These shapes are incredibly complicated to describe in Euclidean terms, yet share an affinity with the so-called "pathological" shapes of modern mathematics, displaying an endless series of motifs within motifs repeated at all scales.

The Koch Curve

One such "pathological" shape is the snowflake curve, devised by
Helge von Koch (1870–1924) in 1904. He defined the curve as the limit
of an infinite sequence of increasingly wrinkly curves. The finished curve
is infinitely long, despite being contained in a finite area. It has no
tangent or smoothness anywhere. Slicing the curve at certain angles
reveals an infinity of Cantor sets lurking within.

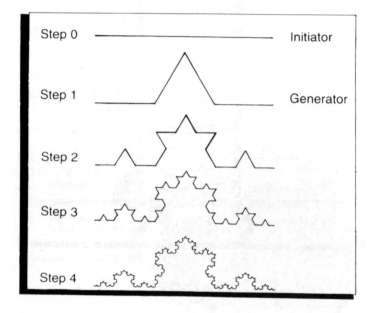

*What Koch didn't realize was that such curves with infinite length would
make ideal models for the shapes of the real world, like coastlines and
arteries.*

Similarity Dimension

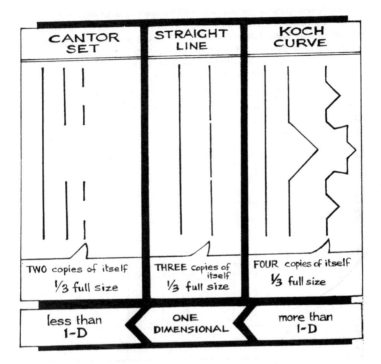

CANTOR SET	STRAIGHT LINE	KOCH CURVE
TWO copies of itself ⅓ full size	THREE copies of itself ⅓ full size	FOUR copies of itself ⅓ full size
less than 1-D	ONE DIMENSIONAL	more than 1-D

A Cantor set contains two one-third-sized copies of itself. A straight line can be split into three one-third-sized copies of itself. A Koch curve consists of four one-third-sized copies of itself. A square is made up of nine one-third-sized copies of itself. In some sense the Cantor set and Koch curve lie on either side of the straight line. The Koch curve is between the line and the square. It takes up more space than the line, but less space than the square. It lies somehow *between* the first and second dimensions! This can be described precisely by the concept of **similarity dimension**, as you can see on the next page.

Similarity and Fractal Dimension

A cube, which is three-dimensional, can be cut into eight (two to the power of three) half-sized cubes. If we know the dimension of an object, powers or exponents allow us to work out how many smaller copies of itself the object contains, of any particular size. An n-dimensional shape is composed of m^n 1/m-sized copies of itself. This suggests a generalization of the concept of dimension, which allows fractional values.

	LINE	SQUARE	CUBE
DIVISION BY 2			
NUMBER of COPIES of ITSELF	2 2^1	4 2^2	8 2^3
DIVISION BY 3			
NUMBER of COPIES of ITSELF	3 3^1	9 3^2	27 3^3

Logarithms are the reverse of exponents. If we know how many smaller copies of itself an object contains, and the relative size, logarithms allow us to calculate the dimension of the object. And it needn't be a whole number.

The Koch curve contains four one-third-sized Koch curves. Its dimension is therefore log4/log3, or about 1.26.

The Cantor set contains two one-third-sized Cantor sets. It has dimension log2/log3, or about 0.63.

In 1919, Felix Hausdorff extended the notion of the similarity dimension to cover all shapes, not just the exactly self-similar. In general, fractal shapes that lie somewhere between dimensions have fractional Hausdorff dimension. Because Hausdorff's definition distinguishes fractal from non-fractal shapes, it is often called the fractal dimension.

The fractal dimension describes the fractal complexity of an object. For example, the coastline of Britain has a fractal dimension of approximately 1.26, about the same as the Koch curve, but slightly less than the outline of a typical cloud (about 1.35). On this scale, a dimension of 1 means totally smooth, while tending towards 2 implies increasing fractal complexity.

Hausdorff's dimension was only a theoretical device until Mandelbrot resurrected it and put it to use. Mandelbrot had the vision to realize that this was the perfect tool to describe the irregularity of nature.

Measuring Fractal Dimension

Although general in scope, the Hausdorff dimension is often difficult to calculate in practice. A simpler way to measure the fractal dimension of a curve is called the box-counting method. Cover the curve with a grid of little squares and count how many squares it passes through. Repeat this process with smaller and smaller squares. In the limit for a fractal curve, the rate at which the proportion of filled squares decreases gives the fractal dimension.

straight line

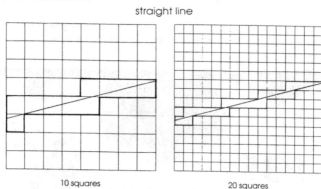

10 squares   20 squares

fractal dimension = $\log(20/10)/\log(2) = 1$

Julia set

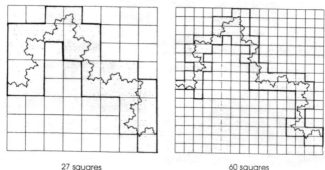

27 squares   60 squares

fractal dimension = $\log(60/27)/\log(2) = 1.152...$

In the case of a straight line, halving the size of the squares requires twice as many squares to cover it. However, for a fractal, more than twice as many squares are needed. A two-dimensional shape requires four times as many squares. A fractal curve is somewhere in between, requiring more than twice, but less than four times, as many.

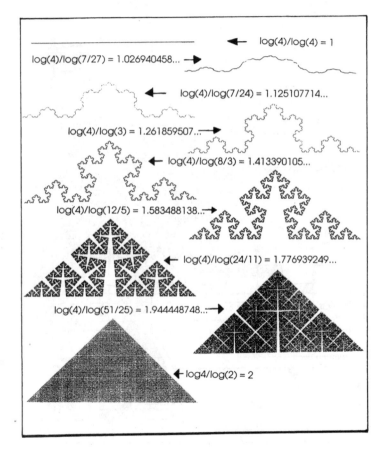

← log(4)/log(4) = 1

log(4)/log(7/27) = 1.026940458... →

← log(4)/log(7/24) = 1.125107714...

log(4)/log(3) = 1.261859507... →

← log(4)/log(8/3) = 1.413390105...

log(4)/log(12/5) = 1.583488138... →

← log(4)/log(24/11) = 1.776939249...

log(4)/log(51/25) = 1.944448748... →

← log4/log(2) = 2

Lewis Richardson

Acting like an 18th-century naturalist, Mandelbrot scoured through forgotten and obscure journals in his quest for insight.

Mandelbrot had struck a rich seam, and he knew it.

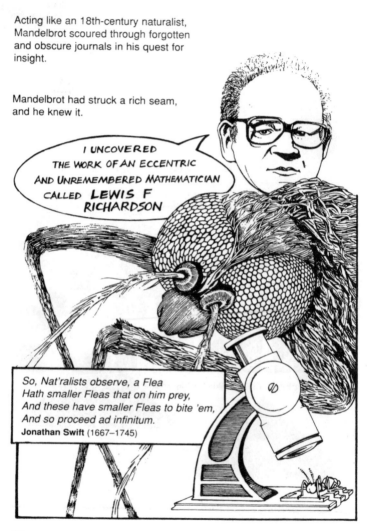

I UNCOVERED THE WORK OF AN ECCENTRIC AND UNREMEMBERED MATHEMATICIAN CALLED **LEWIS F RICHARDSON**

So, Nat'ralists observe, a Flea
Hath smaller Fleas that on him prey,
And these have smaller Fleas to bite 'em,
And so proceed ad infinitum.
Jonathan Swift (1667–1745)

Richardson delighted in asking questions that no one else even considered worth asking. One of his papers, entitled "Does the wind possess a velocity?", anticipated later work by **Edward Lorenz** (b. 1917) and the other founders of chaos theory.

One of this mathematician's great insights was a model of turbulence as a collection of ever-smaller eddies. He conveyed the idea poetically in the style of Swift.

*Big whorls have little whorls,
That feed on their velocity;
And little whorls have lesser whorls,
And so on to viscosity.*

How Long is a Coastline?

What really struck Mandelbrot full force was Richardson's 1961 paper entitled "How long is the coastline of Britain?". What appears to be a simple question of geography transpires, upon further consideration, to expose some of the essential features of fractal geometry. Richardson concluded that the length of the coastline is not well defined.

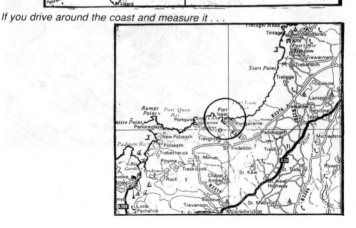

If you drive around the coast and measure it . . .

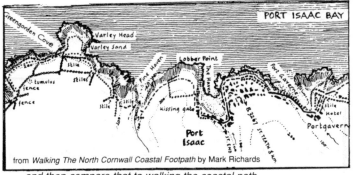

PORT ISAAC BAY

from *Walking The North Cornwall Coastal Footpath* by Mark Richards

. . . and then compare that to walking the coastal path . . .

The measured length of the coastline depends upon the size of your measuring stick.

In the limit, as the measurements become increasingly accurate, the length of the coastline approaches **infinity**.

In other words, the coastline of Britain, and any other natural geological formation, is fractal.

. . . and then stroll it, following every nook and cranny of the beach . . .

. . . you will find that your results do not converge.

They keep getting bigger and bigger.

After an exhaustive analysis of all the cartographical data available to him, Richardson plotted a graph of his results against the logarithm of the size of the measuring stick. Mandelbrot twigged that the slope of Richardson's graph was none other than the Hausdorff dimension of the coastline.

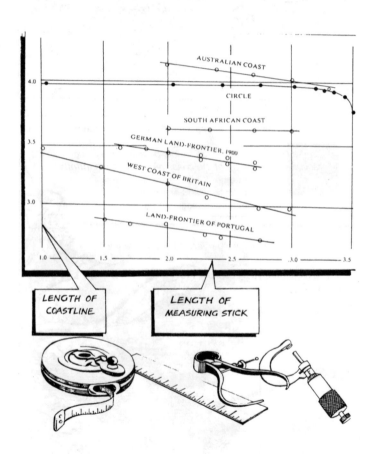

LENGTH OF COASTLINE

LENGTH OF MEASURING STICK

FROM
RICHARDSON'S DATA
I DEDUCED THAT THE COASTLINE
OF BRITAIN HAS A FRACTAL DIMENSION
OF ABOUT 1·26, ABOUT THE SAME
AS THE KOCH CURVE

This is the measure of the wrinkliness of the coastline. So, Richardson's original question can now be rephrased as: "How wrinkly is the coastline of Britain?" The question can now be answered in terms of fractal dimension.

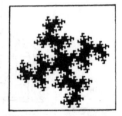

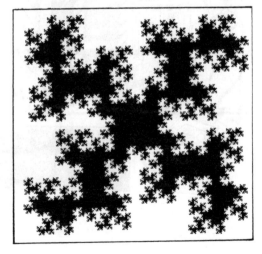

The Sierpinski Gasket

The Polish mathematician **Vaclav Sierpinski** (1882–1969) introduced his fractal in 1916, but the underlying principles had been known to artists for millennia. Early prototypes of the Sierpinski gasket appear on the 12th-century pulpit of the Ravello cathedral, designed by Nicola de Bartolomeo of Foggia, where they have been sketched by the famous artist of "graphic enigmas", **Maurits Escher** (1898–1972).

The Sierpinski gasket is obtained by starting with a filled equilateral triangle, which is then divided into four smaller equilateral triangles, of which the middle one is removed, leaving a triangular hole.

The three remaining filled equilateral triangles are then divided in exactly the same fashion, so that three smaller triangular holes appear.

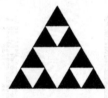

Conceptually we can repeat this process indefinitely, at smaller and smaller scales, reaching, in the limit, Sierpinski's gasket, a shape composed of three copies of itself, each half as big as the whole.

The same thing can be done with squares, pentagons or any other polygons. Similar techniques work with circles. These patterns, made by successively carving three circles out of ever-decreasing circles, show remarkable similarity to Celtic art.

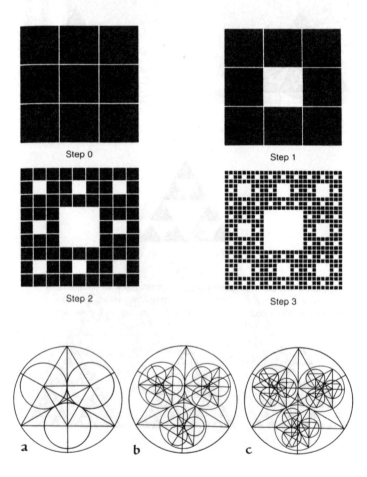

Step 0

Step 1

Step 2

Step 3

a

b

c

Book of Durrow
Ireland, c. 650AD.

The same methods can be adapted to three-dimensional shapes, to produce fractal pyramids and cubes.

The Sierpinski gasket is a product of the mind, an exercise of pure mathematics. Yet remarkably similar patterns are found on sea shells. Fractal shapes often emerge as a result of cellular evolution.

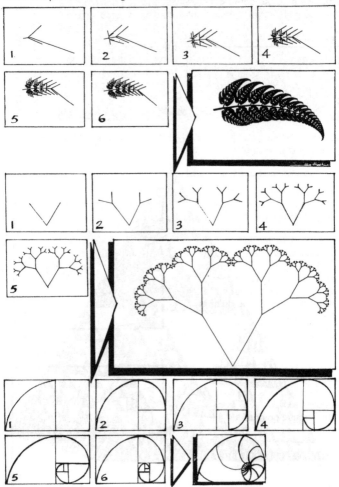

The Chaos Game

In the 1980s Michael Barnsley discovered another way of generating a fractal. It's a bit like dot-to-dot drawings, only you don't join the dots with lines. You just plot point after point according to some simple rules.

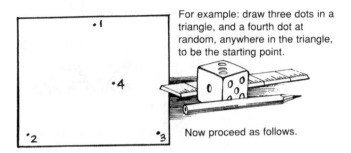

For example: draw three dots in a triangle, and a fourth dot at random, anywhere in the triangle, to be the starting point.

Now proceed as follows.

Step 1. Roll a die.

Step 2. If the top face is 1 or 2, draw another dot half-way from the starting point to the first point. If the top face is 3 or 4, draw another dot half-way from the starting point to the second dot.

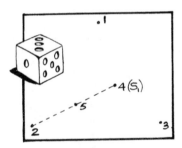

If the top face is 5 or 6, draw another dot half-way from the starting point to the third dot. This new dot then becomes the new starting point, and the whole process is repeated, from Step 1.

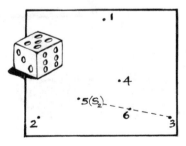

43

After a while, a pattern begins to emerge, a familiar pattern: the Sierpinski triangle.

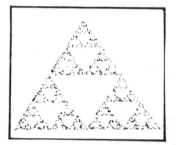 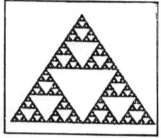

It's not much of a game. With only one player having just one move, the game does not allow for a great deal of choice. After you have chosen your initial point, the future of the game is decided.

As Michael Barnsley discovered, if we had chosen different points, we could have generated a fern instead, or any other fractal – or any shape at all, come to that. Every picture can be encoded as a fractal formula like this.

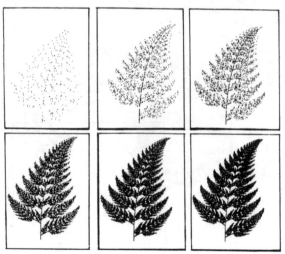

Strange Attractor

What is less obvious is that, in the long run, your initial decision actually makes no noticeable difference at all. What actually happens is that the points are pulled towards what is known as an "attractor" – a **strange** attractor in this case!

Benoît Mandelbrot has questioned the wisdom of this name, preferring "fractal attractor", on the grounds that this type of attractor is actually by far the commonest in nature, so not really that strange after all.

Pascal's Triangle

Another strange thing was the discovery of the Sierpinski triangle inside Pascal's triangle.

Remember school algebra:
$(1 + x)^2 = 1 + 2x + x^2$
$(1 + x)^3 = (1 + x)(1 + x)^2$
$= 1 + 3x + 3x^2 + x^3$
$(1 + x)^4 = 1 + 4x + 6x^2 + 4x^3 + x^4$

Forgetting the x's and just looking at the coefficients, a pattern emerges.

Pascal's Triangle

This graphical representation was known to ancient cultures around the world.

These numbers are fundamental in mathematics.

The numbers in Pascal's triangle represent the number of choosing some number of objects out of some other number of objects.

So, for example, the 6 in the fourth row gives you the number of ways of choosing two identical objects out of four.

possible choices:

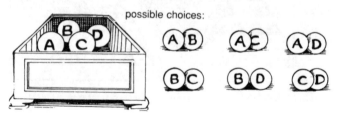

46

What happens if you colour black the odd numbers in the triangle, leaving the even ones white?

A familiar shape appears.

Basins of Attraction

Fractals often appear as the boundary between different zones of attraction. Imagine three fixed magnets, each aligned so as to attract a further magnet which swings above them on a pendulum. The hanging magnet will swing around for a while before coming to rest above one of the fixed ones. Sometimes it will travel almost directly to a stable position of rest. At other times it will oscillate many times from one area to another before finally coming to rest.

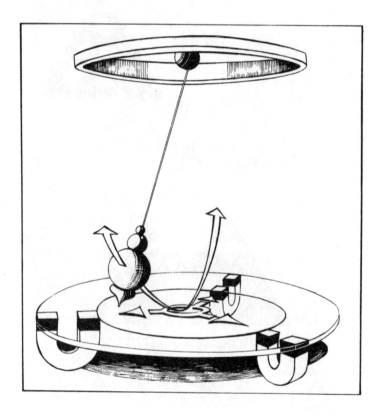

Colouring the zones of attraction black, grey and white leads to the following observation.

Between white and black there is always grey.
Between black and grey there is always white.
Between grey and white there is always black.

In this case, the attraction is physical. The pendulum is literally pulled in to one of the magnets. However, many attractors exist on a more abstract level. When the population of a species reaches a stable level, this state can be thought of as an attractor for the **ecosystem**. There are many other and stranger attractors competing for control of their abstract space.

Poincaré and Non-Linearity

Henri Poincaré, mathematician, physicist and philosopher of science, showed that deep insights into the rather complicated behaviour of dynamical systems can be obtained from quite simple mathematical models.

Henri Poincaré

Shapes like the Koch curve and the Sierpinski triangle are constructed analytically and intentionally to have their strange properties. What Poincaré discovered, at the beginning of the 20th century, was a class of fractals that just emerged spontaneously and unexpectedly from non-linear equations.

More than half a century had to pass, however, before technology had progressed far enough for these fractals to become visible on a computer screen.

Poincaré was looking at the abstract world of mappings, but was limited in how far he could take his research. Years later, doing the dirty work on pocket calculators, the efforts of ecologists and economists to model real systems lead back to these very same mappings.

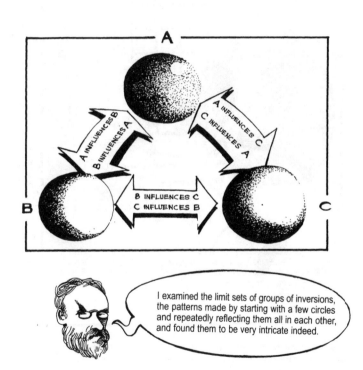

Malthus and Population Growth

The English economist **Thomas Malthus** (1766–1834) observed that
the human population was increasing at an exponential rate, while food
production was only growing linearly.

Extrapolating into the future, I predicted that the
widening gap between demand and available
resources would lead to mass starvation across
the planet.

This population model has a number of weaknesses. In particular, it
assumes a constant rate of population growth, and does not allow for
negative feedback. In reality, once a population has reached a certain
level, the brakes come on and the growth is checked.

Negative Feedback

In the 1840s, Belgian mathematician **Pierre François Verhulst** (1804–49) refined Malthus' model to allow for negative feedback. He assumed that the population of a species in one year was a simple function of the population in the previous year.

Assuming the rate of change to be proportional to distance from maximum population, Verhulst came up with a much more realistic model of population growth. This model predicted that, under the right circumstances, population would stabilize in equilibrium.

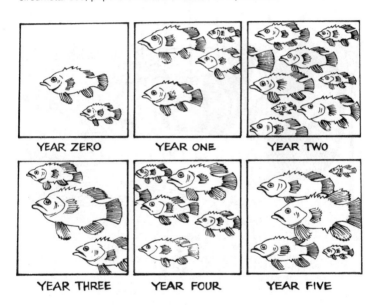

YEAR ZERO   YEAR ONE   YEAR TWO

YEAR THREE   YEAR FOUR   YEAR FIVE

If the population fell below a certain level one year, then it would tend to increase in the next.

While if it rose too high, competition for space and resources would tend to bring it down.

The Logistic Equation

The formula which Verhulst derived is now known as the logistic equation. The revival of interest in this equation in the 1970s led to some of the most beautiful discoveries in the embryonic science that would come to be called chaos theory.

In fact, the study of simple feedback models had scarcely advanced since Verhulst, because without electronic assistance, the calculations involved were far too tedious and cumbersome to perform.

The Verhulst formula itself is very simple, but because you have to repeat the process over and over again, it ends up being very, very complicated.

If x is the population now, then the population next year is given by

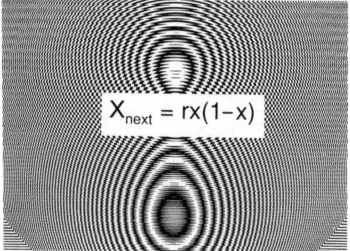

$$X_{next} = rx(1-x)$$

where r is some constant which can be adjusted according to the population being modelled.

As we've pointed out, we can estimate the population of a species in one year as a function of the population level of the year before. If we are interested – and ecologists are – in the long-term behaviour of systems, we have to repeat this formula over and over and over again and see what happens. This process is called **iteration**.

Iteration

Iteration means the repeated application of a rule or a step in an **algorithm**.

Algorithm means the rules for performing complex calculations by a sequence of simpler ones, e.g., digit-by-digit multiplication. Computer programs use algorithms.

It's like a dog chasing its own tail. The output of one operation becomes the input of the other, and so on and on.

> Evolution is Chaos with feedback. God does play dice with the universe, but the dice are loaded. The goal of mathematics and physics now is to find out by what rules the dice were loaded.
>
> Joseph Ford, Georgia Institute of Technology

It's simplest if we take values of x between 0 and 1, so that 1 is the maximum population of the fish pond, and 0 means extinction. Take an arbitrary value for r of 2.6, and so begin.

Suppose $x = 0.2$. Then $1-x = 0.8$, and $x(1-x) = 0.2 \times 0.8 = 0.16$.

Multiply by 2.6 and you get 0.416.

Now repeat the process. Start with $x = 0.416$ and you get 0.6317. The fish are on the increase.

Start with 0.6317 and you get 0.6049. The population falls.

Start with 0.6049 and you get 0.6214. The population rises again.

Then 0.6117, 0.6176, 0.6141, 0.6162, 0.6150, 0.6156, 0.6152, 0.6155, 0.6153, 0.6154, 0.6153, 0.6154, 0.6154, 0.6154.

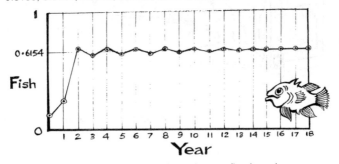

The population rises and falls, but closes in on a fixed number.

Video Feedback

Here is the classic example of iteration. A live camera points at the screen that it feeds, producing an image of an endless tunnel of screens within screens. If each is smaller than the one before it, these images ultimately vanish to a point.

*This point is an **attractor** for the system.*

As Michael Barnsley has shown, if you add another screen, each screen now shows two screens, each showing two screens, etc. The screens are no longer all inside each other: the limit set is much more complicated.

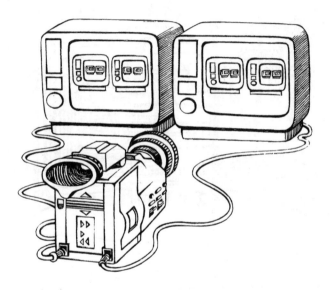

*It is in fact a **strange attractor**, and can take a wide variety of fractal forms, including the Cantor set.*

With more screens, even greater variety is possible. The Sierpinski triangle requires three screens.

Robert May and the Verhulst Model

In the 1970s, **Robert May** (b. 1936) was one of those intrepid ecologists who, in his attempts to unravel the mystery, turned his attention back to the Verhulst model. Mathematicians thought he was mad.

There are many surprises in store for mathematicians, which flow from this simple quadratic iterator.

Robert May uncovered a vast variety of different types of population behaviour following this tack. The simplest of all was **stable equilibrium**. The population would home in on a stable equilibrium and stay at that fixed level, as we've shown with our example on page 55, when we took r = 2.6.

Bifurcation Points

Then, May observed, increasing the system's sensitivity, something strange happens: **oscillation**. Overpopulation in one year is over-compensated for in the next, leading to a lower population, which in turn allows for a boom back to higher levels the year after, and so on.

The pattern then repeats every two years.

Take r = 3.1, and calculate the fish population as we did before. Start with x = 0.2. Then the next value is 3.1 x 0.2 x 0.8 = 0.496. Starting with x = 0.496, we get 0.7750, and then 0.5407, 0.7699, 0.5492, 0.7675, 0.5531, 0.7662, 0.5532, 0.7626, 0.5612, 0.7633, 0.5600, 0.7639, 0.5592, 0.7641, 0.5587, 0.7643, 0.5585, 0.7644, 0.5582, 0.7645, 0.5582, 0.7645. The pattern never settles to a single value, but to **two**. It repeats every **two years**.

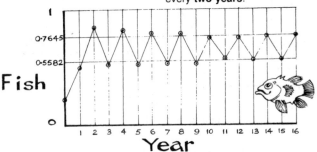

This tendency of stable orbits to split in two at certain critical "bifurcation points" as a parameter in a system changes, is known as **period-doubling**. A single attractor splits in two, and becomes an attractive cycle of period two. This can be represented graphically as a branching line.

The Period-Doubling Cascade

So far we've looked at r = 2.6, which gave a single value of 0.6154, and at r = 3.1, which gave two values: 0.5582 and 0.7645.
Now let's try r = 3.5.
Start as before with x = 0.2.
We get: 0.56, 0.8624, 0.4153, 0.8499, 0.4465, 0.8649, 0.4088, 0.8459, 0.4562, 0.8675, 0.4026, 0.8418, 0.4661, 0.8710, 0.3933, 0.8352, 0.4817, 0.8738, 0.3860, 0.8296, 0.4948, 0.8750, 0.3830, 0.8271, 0.5004, 0.8750, 0.3828, 0.8270, 0.5011, 0.8750, 0.3828, 0.8270, 0.5011.

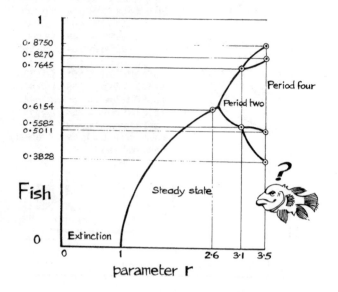

But it doesn't stop there. Increasing the system's response still further results in further bifurcations: the cycle of period two changes abruptly to one of period four. Each branch of the line itself branches in two. This successive splitting in two happens faster and faster, leading rapidly to cycles of period eight, sixteen, thirty-two and so on . . .

The Fig Tree

Like Zeno's arrow, these bifurcations get closer and closer together until they accumulate at one point, known as the Feigenbaum point. At this point, the system converges towards an infinite cycle: it never repeats itself. This is what mathematicians mean by a chaotic orbit. This remarkable pattern is known as "the fig tree". We think you'll see why.

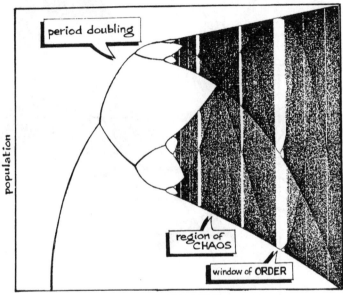

This repeated doubling is known as the **period-doubling cascade**, and it leads to chaos. But the chaos was found to contain within it seeds of order, revealing the existence of windows of stable periodic behaviour carved out of a wall of chaos. The largest of these windows represents a cycle of period three.

Can we show a window of stability? Yes, easily, in the Feigenbaum diagram.

Chaos Theory and Fractals

Tien Yien Li and **James Yorke**'s 1975 paper, "Period three implies chaos", contained the first instance of this usage of chaos in the scientific literature, and led ultimately to the creation of a new science, **chaos theory**.

Li and Yorke's results were startling. They showed that any one-dimensional dynamical system with a cycle of period three also contains cycles of all other periods. **Alexei Sarkowski** had already proved the main result in a stronger form, but, as he wrote in Russian, his work did not receive due attention from the international community. In fact, Sarkowski proved the existence of a magic sequence, which is effectively the order in which different periods arrive on the fig tree.

A new conception of chaos arose. Like relativity, the uncertainty principle and **Kurt Gödel**'s theorem, chaos theory prescribes limits on our knowledge. It says that there are many things we just cannot know.

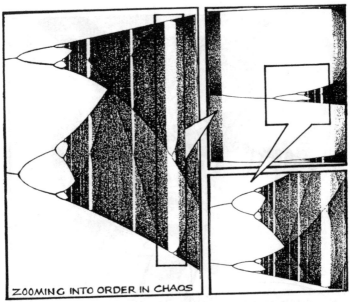

ZOOMING INTO ORDER IN CHAOS

However, the theory has some very positive aspects. It implies that huge changes can be made using a minimal amount of effort. Spacecraft can be slung around the solar system with just a nudge in the right direction. Your life can be radically altered by just the smallest exertion of will at just the right time.

THE SIGNATURE OF CHAOS IS THE FRACTAL ATTRACTOR — FRACTALS ARE THE PATTERNS OF CHAOS

The Feigenbaum Constant

Pursuing the themes of iteration and bifurcation, in 1977 **Mitchell Feigenbaum** showed that the ratio of distances between successive bifurcations converged rapidly to a constant, 4.669201660910 . . .

Powerful vindication of the importance of this result came with the derivation of exactly this same value from various physical laboratory experiments. The universality of the period-doubling cascade in mathematics led some theorists to predict its occurrence in nature. Still, many were astonished when the cascade started to appear in real physical systems as diverse as acoustic feedback and dripping taps. Period-doubling is a general principle of nature.

Physical instances were revealed in labs around the world. And when the rate of successive doubling was measured, experiment after experiment showed remarkable numerical agreement with Feigenbaum's number, 4.6692016 . . .

Feigenbaum's number is a truly universal constant, as fundamental as π (ratio of a circle's circumference to its diameter: 3.141592654. . .). It applies just as much in the real world as it does in computer simulations.

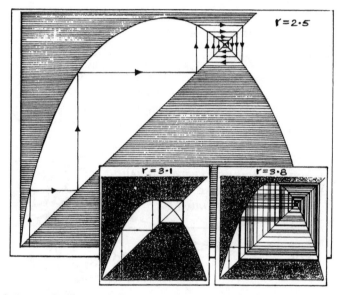

It appears in all natural phenomena governed by a one-humped feedback process.

In the Mandelbrot set, it is the ratio of radii of successive circles on the real line.

Real Numbers, Other Numbers

As Ian Stewart says: "Feigenbaum's discovery of universality is a two-edged sword. It makes it relatively easy to test a particular class of chaotic models by experiment; but it doesn't distinguish between the different models in that class."

Feigenbaum's discovery turned out to be just a small part of a much bigger picture. Feigenbaum had looked only at iterations of the logistic equation for real numbers.

THAT IS, THE EVERYDAY NUMBERS WE USE FOR COUNTING AND QUANTIFYING THINGS LIKE POPULATION AND PRICES

BUT MATHEMATICALLY THESE NUMBERS ARE A SPECIAL CASE OF A MORE GENERAL TYPE OF NUMBER

real

imaginary

When Benoît Mandelbrot extended Feigenbaum's work to this wider domain, a pattern of astounding beauty was revealed.

Complex Numbers

Complex numbers are generated by allowing for there to be square roots of minus numbers. This is not something we encounter in our day-to-day lives. There are no minus things out there in the world we live in. Unless you count dark matter! And you can't *really* calculate the square root of a minus entity. Can you?

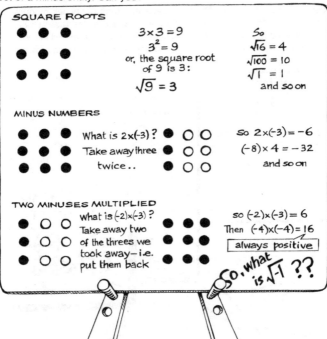

SQUARE ROOTS

$3 \times 3 = 9$
$3^2 = 9$
or, the square root of 9 is 3:
$\sqrt{9} = 3$

So
$\sqrt{16} = 4$
$\sqrt{100} = 10$
$\sqrt{1} = 1$
and so on

MINUS NUMBERS

What is $2 \times (-3)$?
Take away three
twice..

So $2 \times (-3) = -6$
$(-8) \times 4 = -32$
and so on

TWO MINUSES MULTIPLIED

What is $(-2) \times (-3)$?
Take away two of the threes we took away – i.e. put them back

so $(-2) \times (-3) = 6$
Then $(-4) \times (-4) = 16$
always positive

So, what is $\sqrt{-1}$??

Well, in maths and in the mind you can, if you allow for imaginary numbers! The square root of minus one is symbolised by the letter *i*. The complex numbers were conceived as tools for solving cubic equations, but at no extra cost provide solutions to any equation of any order.

They express deeper patterns of reality than the so-called "real" numbers alone. Complex numbers are central to the formulation of quantum mechanics and the description of oscillating systems. Let's now see how complex numbers relate to fractals.

The Complex Plane

In 1685, the English mathematician **John Wallis** (1616–1703) hit upon the idea of representing complex numbers graphically, in a diagram. The imaginary numbers are strung out at right angles to the line of real numbers, also known as the real line. If we portray real numbers as lying on an east/west axis, then imaginary numbers lie on the perpendicular axis, through the point zero, and run from north to south. This creates a cross – a co-ordinate system, where all the real numbers are placed on one axis and all the imaginary numbers on the other, with a general point in the plane consisting of a real part and an imaginary part.

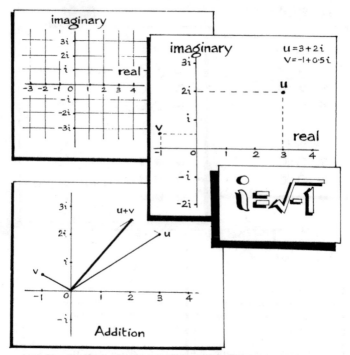

Arithmetic operations on complex numbers correspond to geometrical transformations of the complex plane. Addition is equivalent to translation; multiplication to scaled rotation.

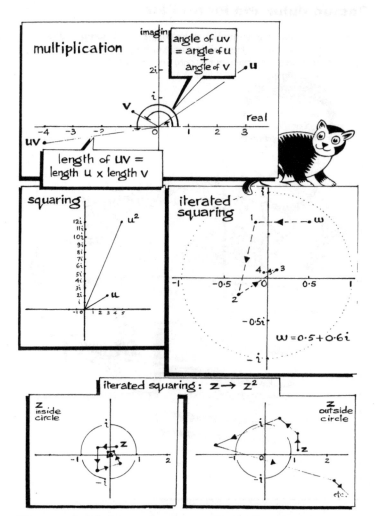

When equations in the complex plane are iterated through a computer, remarkable forms like the Mandelbrot set appear.

Gaston Julia and Pierre Fatou

During the First World War, the French mathematicians **Gaston Julia** (1893–1978), a student of Henri Poincaré, and **Pierre Fatou** (1878–1929) studied the rational mappings of the complex plane.

A transformation, or mapping of the plane, is a rule which, given any point in the plane, provides another.

It can be thought of as acting simultaneously on the entire plane . . .

. . . picking it up, stretching it, spinning it, or twisting it, then laying it down flat again.

Julia and Fatou looked in particular too at the process of iteration. Their work remained largely unknown, even to most mathematicians, because without modern computer graphics it was almost impossible to communicate their subtle ideas.

Self-similarity, for example, was well known to Julia and Fatou. The mappings Julia studied were discrete, yet various analogies with real continuous dynamical systems arose naturally.

Repellors created basins of attraction, much like falling rain has to find its way into one river or another. The boundary between the catchment areas was made up of repelling points, which pushed neighbouring points away. These boundaries turned out to be very complicated. They are now known as Julia sets.

Julia and Fatou never actually saw a Julia set. The first visualization appeared in a paper published in 1925. This was the closest they ever came to seeing the full flowering of their work.

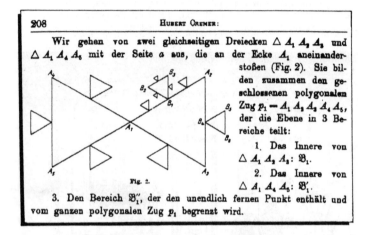

Wir gehen von zwei gleichseitigen Dreiecken $\triangle A_1 A_2 A_3$ und $\triangle A_1 A_4 A_5$ mit der Seite a aus, die an der Ecke A_1 aneinanderstoßen (Fig. 2). Sie bilden zusammen den geschlossenen polygonalen Zug $p_1 = A_1 A_2 A_3 A_4 A_5$, der die Ebene in 3 Bereiche teilt:

1. Das Innere von $\triangle A_1 A_2 A_3$: $\mathfrak{B}_1$.

2. Das Innere von $\triangle A_1 A_4 A_5$: $\mathfrak{B}_1'$.

Fig. 2.

3. Den Bereich $\mathfrak{B}_1''$, der den unendlich fernen Punkt enthält und vom ganzen polygonalen Zug p_1 begrenzt wird.

It was not to be until the advent of modern computers that Julia sets could be seen in their full glory, and with all their detail exposed.

Hidden behind the apparent diversity of the Julia sets, a powerful unifying principle was waiting to be discovered. A key to untold mysteries. A sleeping dragon. Only one person was radical enough to find it.

  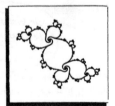

Benoît Mandelbrot

Mandelbrot was born in Warsaw in 1924 to a relatively wealthy family. His father was a successful clothing wholesaler and his mother a well-respected dentist. They were Lithuanian Jews. When Benoît was twelve, his family had the foresight to quit Poland. Persecution loomed dark and heavy over them. Paris was their destination.

They had relatives and friends in Paris, who were prepared to help them find work and accommodation.

Uncle Szolem

Among their contacts was Benoît's Uncle Szolem Mandelbrot – a mathematician and his father's younger brother. Although this move was difficult and painful, it did have beneficial and lasting repercussions for Benoît. His uncle took him under his wing. As Mandelbrot recalls: "That mathematics is a living organism was revealed to me when I was a child by my uncle Szolem. My uncle and my father, you can say, fought for my soul."

A Practical Education

When Paris fell to Nazi Germany in 1940, the Mandelbrots fled south. Benoît became an apprentice tool-maker. His schooling was irregular and discontinuous. He never actually learned the alphabet, nor multiplication past the fives times table! His practical education, though, opened his eyes to forms occurring in the natural world.

Bare trees in winter look like the estuaries of rivers and like anatomical drawings of the human circulatory system.

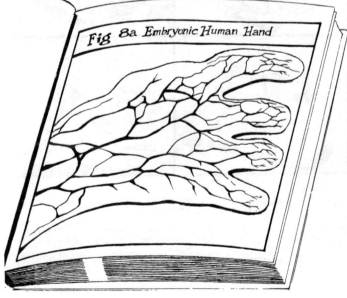

Fig 8a Embryonic Human Hand

The Shape of Things

The cauliflower fascinated him particularly.

He noticed that when you break off a branch from the cauliflower, the small piece looks just like the whole thing.

Then, Benoît discovered, he could continue to break off smaller and smaller branches of the cauliflower and that, up to a certain point, they continued to look like smaller and smaller versions of the whole vegetable.

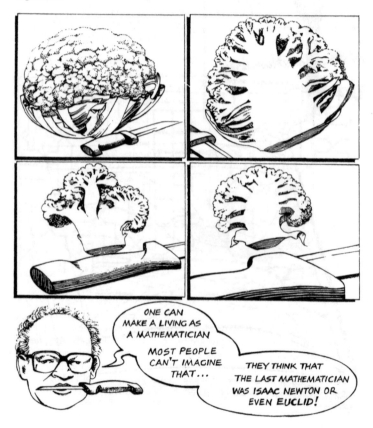

ONE CAN MAKE A LIVING AS A MATHEMATICIAN

MOST PEOPLE CAN'T IMAGINE THAT...

THEY THINK THAT THE LAST MATHEMATICIAN WAS ISAAC NEWTON OR EVEN EUCLID!

Mandelbrot realized early on that he would be a mathematician. He also knew from his Uncle Szolem that mathematics was a real occupation.

Back in Paris after the war, Mandelbrot sat – and passed – the arduous French university admissions examinations with absolutely no preparation. On the mathematical sections of the test – exercises in formal algebra and integrated analysis – he managed to hide his lack of training with the help of his geometrical intuition.

He realized that he had the ability to see the "shape" of an analytical problem in his mind's eye.

Given a shape, he could find ways of transforming it, altering symmetries, creating harmony. Often these transformations led directly to a solution in physics and chemistry. But where he could not employ geometry, he did not fare so well. Sitting these examination papers brought home again to Benoît that there was a "mathematics of the eye", that visualization of a problem was as valid a method as any for finding a solution. Amazingly, he found himself utterly alone with this conjecture.

The Wiles of Bourbaki

The teaching of mathematics in France was dominated by a handful of dogmatic mathematicians hiding behind the pseudonym "Bourbaki", named after the 19th-century French general, Nicholas Bourbaki. It was an inside joke. The name appealed to the members because it sounded strange and attractive. Although Bourbaki met in secret, they influenced mathematical thinking, not just in France, but throughout Europe.

Pictures were considered ephemeral and not fit for the rarefied heights of pure mathematics. Pictures could lead a mathematician astray. Mandelbrot was unable to go along with this kind of thinking. Bourbaki represented everything he abhorred and was struggling hard to avoid. He wanted his maths – his geometry – to explain the "real" world. He yearned for a true description of nature and her processes.

On the Run Again

He married and left France for the USA to escape the ice-cold grip of Bourbaki. It's conceivable that without the insufferable formalism of Bourbaki to kick against, Mandelbrot's restless spirit would not have been released and that we would still be waiting for the Mandelbrot set and fractal geometry!

The stifling hold on mathematical imagination pushed Mandelbrot away from academia to work at IBM in New York.

HERE I WOULD HAVE THE INTELLECTUAL FREEDOM NO UNIVERSITY COULD GRANT ME

The More the Merrier

IBM gave Mandelbrot the funding, the facilities, a research team that included Dr Richard Voss, and the mental space in which to work. The powers-that-be at IBM, it must be said, had vision – unlike the reactionary management of the mainstream academic world. Looking back, Mandelbrot observes . . .

Emerging Patterns

Drawing his inspiration from mathematical mavericks like Richardson, as well as papers rescued from other mathematicians' wastebaskets, Benoît Mandelbrot formulated the new science of fractal geometry.

From the height of the River Nile to the distribution of craters on the moon, wherever he looked he found the same patterns emerging.

Shapes like the Cantor set and the Koch curve, far from being exceptional, are in fact ubiquitous in the physical world.

Sketch of the mathematical simulation of crater patterns on the left.

$A = \frac{K^{(2-D)}}{e}$

Fractals in Practice

Variations on the Cantor set occur in everything from frequencies of words and letters in language to noise on telephone lines, while the Koch curve serves as a model for real coastlines. As Mandelbrot wrote in the introduction to his groundbreaking book, *The Fractal Geometry of Nature* (1977) . . .

The number of distinct scales of length of natural patterns is for all practical purposes infinite.

Fractal forgery of an island

Fractal geometry is indeed very much a geometry of the practical, of the real nuts-and-bolts world.

Bad Noise

Right at the beginning of Mandelbrot's career at IBM, he tackled a practical issue which directly involved and concerned his employers. Inside the company, data was being lost or corrupted when passing between computers by random noise bursts, which they could not get rid of or predict.

He applied his own mathematics and came at the problem in a completely new way.

Fractal Errors

Mandelbrot's solution made no sense to IBM's engineers, but they had to accept that the mathematics did predict that it was just not possible to compute the average rate of errors over any period of time.

IBM accepted the logic in good faith and rebuilt their system to cope with the errors. They introduced a level of redundancy into the system, which would cancel out the interference. The maths was right. Mandelbrot could explain his observations mathematically.

A General Principle Emerges

This is the principle behind non-linear dynamics. Mandelbrot was beginning to grasp, and explain mathematically, the way things work in the real world. He sensed the underlying shape of things – the hidden order.

The Simplest Possible Transformation

Mandelbrot became interested in the Julia set of the simplest possible transformation: $z \rightarrow z^2 + c$. This formula gives a rule for getting one complex number from another, or in other words, mapping the complex plane onto itself.

The effect of this mapping is to cut the plane and wrap it around itself while stretching it away from the unit circle.

Mandelbrot called the Julia sets generated by this formula "self-squared dragons".

The Julia sets for this mapping depend only on the value of the parameter c. When c is small, they are simple loops, like wrinkled circles. For large values of c, the fractal consists of innumerably many discrete points, spread out and dust-like.

Two Types of Julia Sets

Generally, the Julia sets can be divided into two main types, two varieties. They can be either wholly **disconnected** and dust-like, or wholly **connected**. In the former case, the sets are topologically the same as the classic Cantor set.

With the connected Julia sets, on the other hand, each consists of a succession of lines: sometimes a single closed curve; sometimes loops within loops within loops; and, from time to time, a dendrite.

On the boundary between these two regions are dendrite Julia sets, made up completely of continuously sub-branching lines, which are only just connected, since the removal of any point from them would split them in two.

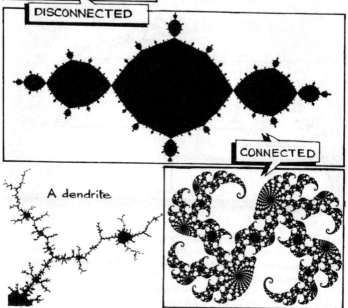

DISCONNECTED

CONNECTED

A dendrite

He and his team spent ages churning out Julia sets.
"It was extraordinary great fun!"

A Map of Julia Sets

In 1980, Mandelbrot hit upon the fruitful idea of making a map of this behaviour. He coloured a point on the map black if a Julia set from that location was connected, and left it white if the Julia set was disconnected. This is how the Mandelbrot set is defined: the Mandelbrot set is the set of points c for which the Julia set of $z \rightarrow z^2 + c$ is connected.

If c is in the Mandelbrot set, the Julia set is connected.

If c is outside the Mandelbrot set, the Julia set is disconnected.

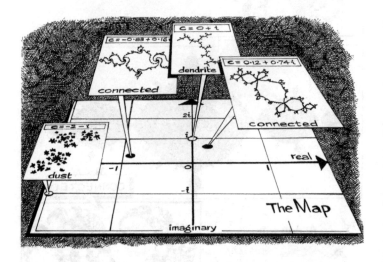

At first glance this seems a rather complicated problem, but Julia had devised a trick for finding out whether or not a Julia set is connected without actually constructing the set itself. If the orbit of the starting point is examined and found to go to infinity, then the set is disconnected; otherwise it is connected. With this insight, a simple computer program can find out to which class a Julia set belongs.

A Whole New World

When the first picture of the set rolled off the printer, Mandelbrot and his colleagues' first reaction was that there must be some mistake, a fault in the program perhaps. The picture was extremely strange and unexpected.

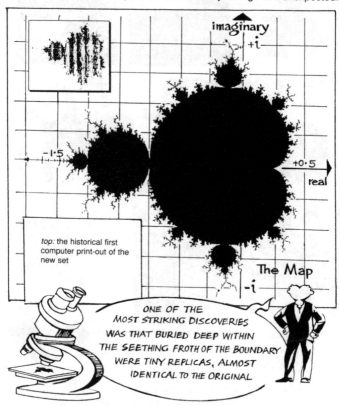

top: the historical first computer print-out of the new set

The Map

ONE OF THE MOST STRIKING DISCOVERIES WAS THAT BURIED DEEP WITHIN THE SEETHING FROTH OF THE BOUNDARY WERE TINY REPLICAS, ALMOST IDENTICAL TO THE ORIGINAL

Over a period of weeks, working late into the night in the basement of a laboratory in Harvard University, Mandelbrot and his assistant explored the astonishing new world they had discovered. By feeding new co-ordinates into their program, they made successively deeper zooms into the boundary of the set.

Heinz-Otto Peitgen

THE M-SET
CONVERGES
TOWARDS
JULIA SETS

Zooming further into these baby M-sets, they encountered further variations of these very same patterns with added frills and embellishments. These baby M-sets are like a key, revealing the existence of a whole new world of embedded Julia sets.

These consist of an infinite number of scaled-down copies of one Julia set, strung on spiral strands in the general shape of another.

The most interesting Julia sets are the ones which are only just connected. These lie on the border between the connected and the dust-like sets – that is, on the boundary of the Mandelbrot set.

Likewise, the Mandelbrot set is most interesting at the boundary. The Chinese mathematician **Tan Lei** proved that the M-set is asymptotically similar to Julia sets near any point on its boundary.

ALL THE PATTERNS FOUND IN SUCH JULIA SETS ARE FOUND ON THE EDGE OF THE M-SET

comparing enlargement of M and Julia set at c = –0.745429 + 0.113008

The more you zoom into the Mandelbrot set, the closer it resembles some particular Julia set. For this reason, the M-set has been called a pictorial index of Julia sets.

Simple Rules, Complex Behaviour

Complex phenomena do not necessarily require complex explanations. This is the essence of chaos theory, beautifully conveyed in the Lorenz attractor.

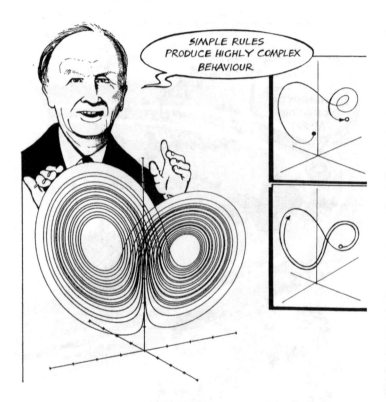

A very simple formula like Newton's law, which is just a few symbols, can explain the motion of the planets around the sun and many, many other things to the 50th decimal!

Newton's Law and Chaos

Although Newton's law describes the periodic orbits of the planets in the solar system, it also implies that the system is in fact chaotic – only stable in the short-term. In this case, millions of years.

The asteroid belt that lies between Mars and Jupiter is clear evidence of the chaos implied by Newton's law. Saturn's rings display a fractal structure akin to the Cantor set, with gaps in critical regions which correspond to unstable orbits.

IT IS IMPOSSIBLE TO PREDICT THE LONG-TERM BEHAVIOUR OF ANY SYSTEM OF THREE OR MORE BODIES

Let's Admit – It is Complicated . . .

The Mandelbrot set is used as evidence for mathematical realism. It is so complicated, the argument goes, that no one could possibly have invented it. This is precisely what the mathematician **Roger Penrose** is saying when he emphasizes the reality of the Mandelbrot set.

The Mandelbrot set is so intricate that no computer program can determine whether or not a general point belongs to it. In some sense, it is not computable.

The further you zoom into the Mandelbrot set, the more intricate it becomes. **Mitsuhiro Shishikura** proved in 1991 that the boundary of the Mandelbrot set has fractal dimension 2.

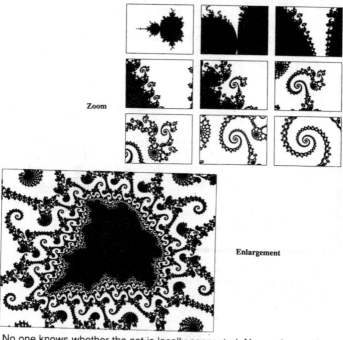

Zoom

Enlargement

No one knows whether the set is locally connected. No one knows the exact area of the Mandelbrot set, though it is known to be around 1.50659177. . . Massive computational power has been expended on this problem, though it could take a while longer before anyone actually finds a use for the result! Nevertheless, the fact that the Mandelbrot set exists and that it has an area is enough for mathematicians of the Everest school to attempt the challenge – because it's there. In the absence of any better ideas, they continue to chop the M-set up into ever smaller boxes, counting them all up to arrive at ever closer estimates. Convergence is very slow, and it has been suggested that maybe the boundary of the M-set has positive area, though no-one knows the answer to this either.

Phase Transitions

We know water in three states: ice, liquid and vapour. The change from one state to another happens abruptly, at certain precise temperatures – 0 and 100 degrees C. This is the essence of phase transitions: discontinuous jumps in a system's behaviour as a parameter crosses critical thresholds. The dynamics of phase transitions can now be explored with computer models. Patterns have been found that are common to all phase transitions.

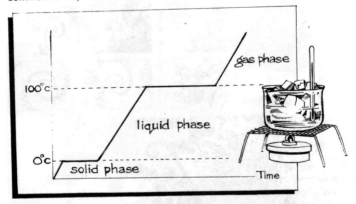

Two German mathematicians, **Heinz-Otto Peitgen** and **Peter H. Richter**, were investigating a model of magnetic phase transitions, and discovered a whole new family of fractals. And yet, deep within the fractal bubbles . . .

Hazy Areas of Calculation

Isaac Newton's method for solving equations by successively better guesses was used by mathematicians for centuries before **Arthur Cayley** (1821–95) noticed that between different solutions lay a highly complex grey area.

A CENTURY LATER COMPUTER GRAPHICS REVEALED THAT THIS HAZY AREA WAS ALSO FRACTAL

WITHIN IT WERE COPIES OF JULIA SETS AND THEIR ORTHOGONAL COUSIN THE **M-SET**, IN PERFECT DETAIL

The recurrence of the Mandelbrot set in a wide range of different fractals is due to the phenomenon of universality. Wide classes of different systems have essentially similar attractors. This was discovered in the one-dimensional case by Mitchell Feigenbaum in 1978, using just a pocket calculator and plenty of stimulants!

The Mathematics of Wrinkles

The M-set is wrinkly and crinkly, just like the natural world. Before fractal geometry there were no tools to describe this aspect of the world we experience.

Climbers are often misled by the lack of distinguishing features which can make a peak appear to be just over the next rise, only to find a vast chasm between them and the top.

Any part of a mountain resembles the whole.

A frond of a fern looks like a whole fern.

This is the fundamental characteristic behind the growth of all complex organisms. The same forms recur in many different circumstances, in different materials both organic and inorganic, on a vast range of scales. A small part of our circulatory system looks like the whole. It looks like a tree, like an estuary, like a stream-bed.

100

Natural Templates

Nature finds the same solution to many different problems, like how to drain water from the land into the oceans, and how to get blood from our hearts to our fingertips and back again. And the templates that nature uses are fractals. Clouds look the same at all scales. It is impossible to determine the size of a cloud from a photograph of it.

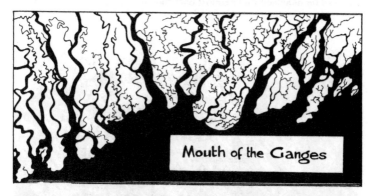

Mouth of the Ganges

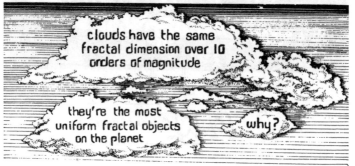

clouds have the same fractal dimension over 10 orders of magnitude

they're the most uniform fractal objects on the planet

why?

Clouds are formed by the condensation of tiny droplets of water, which occurs on a fairly random basis under suitable conditions. However, once these are formed, they tend to attract more tiny droplets at certain points around them. This creates the necessary conditions for a fractal.

Forest Fires: the Fractal Boundary

Imagine a plantation of evenly spaced trees on a very hot, dry day. As the temperature soars, the odd leaf or twig ignites, sending a whole tree up in flames. This is an essentially random process – the factors involved are beyond our powers of prediction. But once a tree is in flames, the fire easily spreads to neighbouring trees, and this process can now be modelled with iterative techniques.

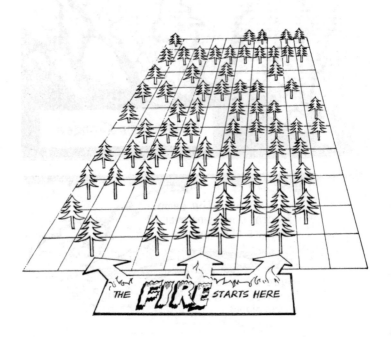

THE FIRE STARTS HERE

*We can't say anything about which tree will go up first, but we can say how the fire will spread. The fire will have a **fractal boundary**.*

More Phase Transitions

This principle applies just as much to the spread of infectious diseases or magnetic polarity as to forest fires. It is essentially the same process unfolding.

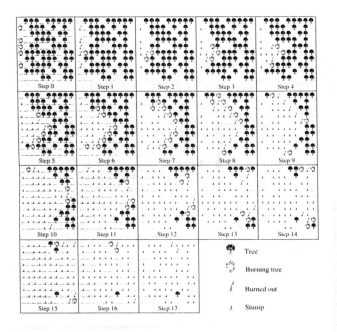

*What we find is highly sensitive dependence to initial conditions, producing **phase transitions**.*

Once a critical threshold is passed, the fire spreads outwards, the disease becomes an epidemic, the material magnetic.

Electro-plated Fractals

In 1983, **Thomas Witten** and **Leonard Sander** found a new way of looking at how deposits build up in the electro-plating process. Their work led to a new model called "diffuse limited aggregation", or DLA. Computer simulations using their model are indistinguishable from the real thing.

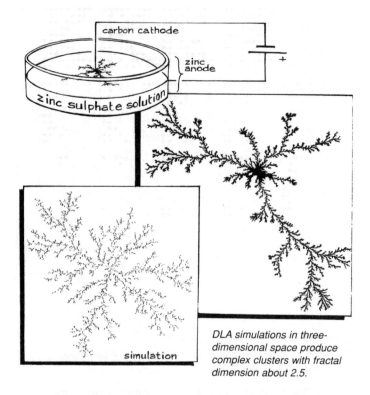

DLA simulations in three-dimensional space produce complex clusters with fractal dimension about 2.5.

They remind us of seaweed and the patterns produced by oil in water, called **viscous fingering**.

Viscous Fingering

In the classic Hele-Shaw experiment, water is fed through a small hole in the middle of a flat layer of oil sandwiched between two sheets of glass.

At first the water spreads out uniformly, but fairly soon the increasing straight interface between the two fluids becomes unstable and breaks into fjord-like fronds, which fan outwards in a fractal manner reminiscent of coral or plant growth.

Viscous fingering strongly resembles diffuse limited aggregation (DLA).

Both have a fractal dimension around 1.7, and there is growing evidence that they are mathematically related.

Growth by Aggregation

Coral grows outward by aggregation. That is, successive deposits of material attach to an outward-growing surface. They look very much like trees, which grow outwards by branching from the inside. This is similar to snow flakes forming, and to DLA patterns.

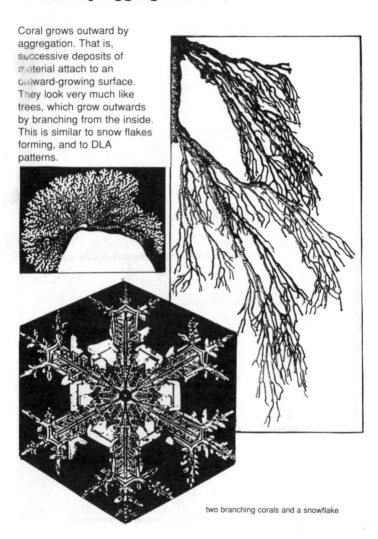

two branching corals and a snowflake

Brownian Motion and the Quantum World

Brownian motion, the movement of a tiny smoke particle due to the constant bombardment of millions of invisible air molecules, traces a fractal path with dimension close to 2.

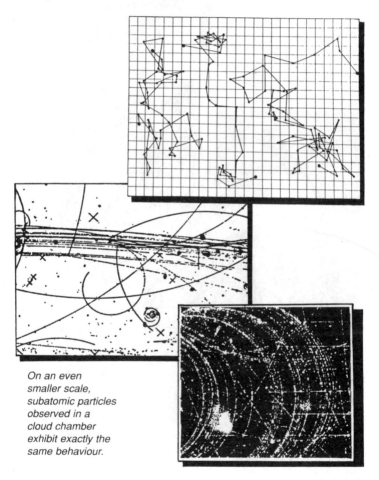

On an even smaller scale, subatomic particles observed in a cloud chamber exhibit exactly the same behaviour.

Dimensional Magic

We are fractal. Our lungs, our circulatory system, our brains are like trees. They are fractal structures.

Fractal geometry allows bounded curves of infinite length, and closed surfaces with an infinite area. It even allows curves with positive volume, and arbitrarily large groups of shapes with exactly the same boundary. This is exactly how our lungs manage to maximize their surface area.

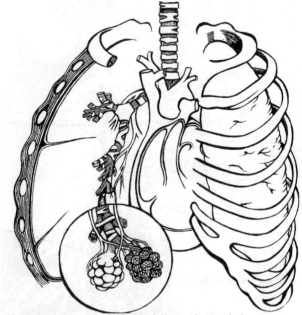

Most natural objects – and that includes us human beings – are composed of many different types of fractals woven into each other, each with parts which have different fractal dimensions. For example, the bronchial tubes in the human lung have one fractal dimension for the first seven generations of branching, and a different fractal dimension from there on in.

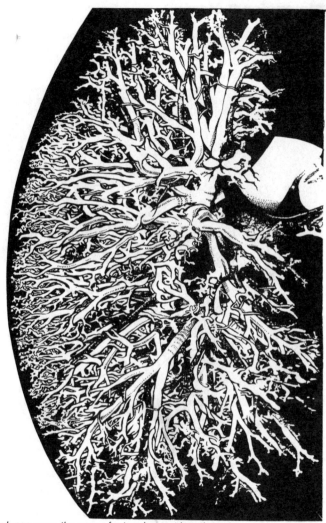

Our lungs cram the area of a tennis court into the volume of just a few tennis balls.

The Three-Quarter Power Law

Fractal geometry has revealed some remarkable insights into a ubiquitous and mysterious "three-quarter power law". This particular power law models the way that one structure relates to and interacts with another. It is based on the cube of the fourth root.

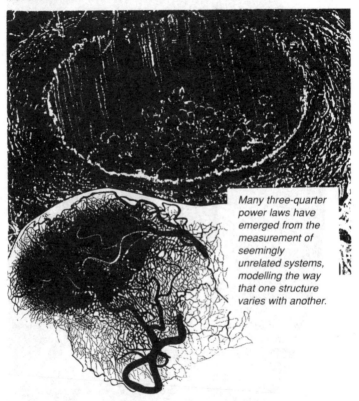

Many three-quarter power laws have emerged from the measurement of seemingly unrelated systems, modelling the way that one structure varies with another.

For a long time now, physiologists have had an empirical understanding of how much blood flows through our circulatory system, and how this relates to the physical size of the vessels that carry it. Research employing fractal rules has revealed a three-quarter power law even in the circulatory system.

Our arteries, which account for just 3 per cent of our bodies by volume, can reach every cell in our bodies with nutrients.

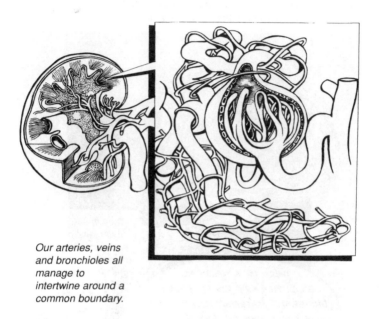

Our arteries, veins and bronchioles all manage to intertwine around a common boundary.

The arteries that deliver the blood, and the veins that take it away, need to share a common interface with the surface of the lungs, in order to aerate the blood. The arteries must provide every cell in our bodies with nutrients, using the minimum amount of blood.

The kidneys, the liver, the pancreas are all organs constructed along self-similar fractal rules. So too is the most remarkable organ of all those that we know on this planet – the **human brain**.

The Mysterious Brain

One thing we can say with certainty about the brain is that it is a very fractal piece of kit! It has an obvious fractal structure. You have only to look at it to see that. It is very crinkled and wrinkled and highly convoluted, as it folds back and back on itself.

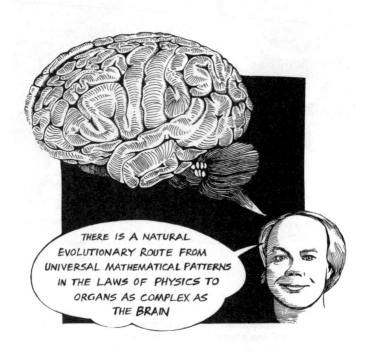

THERE IS A NATURAL EVOLUTIONARY ROUTE FROM UNIVERSAL MATHEMATICAL PATTERNS IN THE LAWS OF PHYSICS TO ORGANS AS COMPLEX AS THE BRAIN

It is deeply ironic that this remarkable organ, which is the seat of the mind, and which either created or discovered (we don't know which) the mathematical rules on which it and the entire universe turns, cannot explain or understand its own functioning.

Knowing the Mind of God?

$$S_{bh} = \frac{A}{4}\frac{kc^3}{G\hbar}$$

OUR VERY AWARENESS — THE AWARENESS THAT CONSTRUCTS AND ANALYSES FRACTALS AND EVERYTHING ELSE — CONTINUES TO REMAIN A MYSTERY TO ITSELF

If we understood how our brains worked, would we not have achieved those dizzying heights conjured up by **Stephen Hawking** in *A Brief History of Time* – would we not, then, "know the mind of God"?

Understanding how our brains function is probably the greatest challenge facing the scientific community at this time. Fractal geometry is at the leading edge of research in this area.

Fractals and Medical Research

All aspects of nature follow mathematical rules and involve some roughness and a lot of irregularity. For example, complex protein surfaces fold up and wrinkle around towards three-dimensional space in a dimension that is around 2.4. Antibodies bind to a virus through their compatibility with the specific fractal dimension of the surface of the cell with which they intend to react.

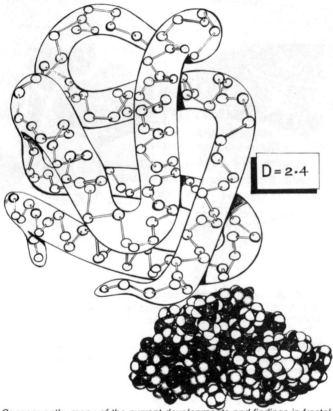

$D = 2 \cdot 4$

*Consequently, many of the current developments and findings in fractal geometry are in work with **surfaces**.*

Viruses and Bacteria

The receptor molecules on the surfaces of all viruses and bacteria are fractal. Their positioning techniques, the methods they use to determine the chemistry of the body they have invaded and how they will interfere with that body's chemistry, and their binding functions, emerge mathematically by way of the deterministic rules of fractal geometry.

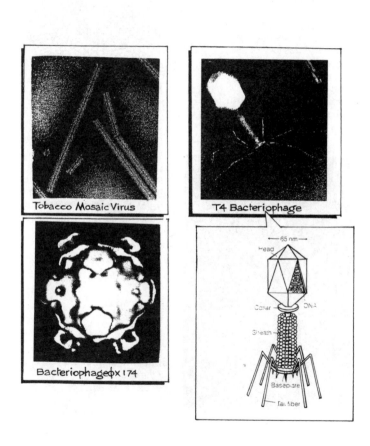

Tobacco Mosaic Virus

T4 Bacteriophage

Bacteriophage φx 174

←— 65 nm —→
Head
Collar
DNA
Sheath
Baseplate
Tail fiber

AIDS

The dynamics of the AIDS virus in the human body has been modelled with fractal geometry. Fractal geometry provides the answer to the long-standing puzzle surrounding the unusually long incubation period of the AIDS virus. Many patients remain HIV positive for as long as ten years before the virus decides to kick in. The onset of the full-blown disease does reveal itself in the body.

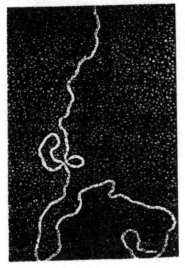

As the immune system begins to fall apart, the AIDS virus starts to behave chaotically. Studies of the virus at this stage have revealed significant changes in the fractal structure.

*Fractal geometry unravels the **structural differences** that occur at the end of the incubation period of the virus.*

Detecting Cancer

The surface structures of cancer cells are crinkly and wrinkly. These convoluted structures display fractal properties which vary markedly during the different stages of the cancer cell's growth.

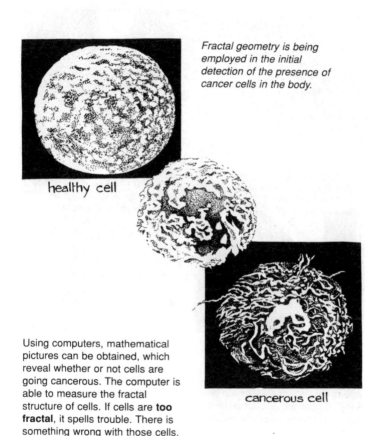

Fractal geometry is being employed in the initial detection of the presence of cancer cells in the body.

healthy cell

cancerous cell

Using computers, mathematical pictures can be obtained, which reveal whether or not cells are going cancerous. The computer is able to measure the fractal structure of cells. If cells are **too fractal**, it spells trouble. There is something wrong with those cells.

Women at Risk

The fractal dimension of cancerous material is higher than that of healthy cells. **Alan Penn**, who is Adjunct Professor of Mathematics and Engineering at The George Washington University, describes his work in this area.

Clinical application of MRI has been hampered by difficulty in determining which masses are benign and which are malignant. Research has focused on developing robust fractal dimension estimates which will improve discrimination between benign and malignant breast masses.

Bubbly Bones and Breaks

Bone fractures are
fractal.

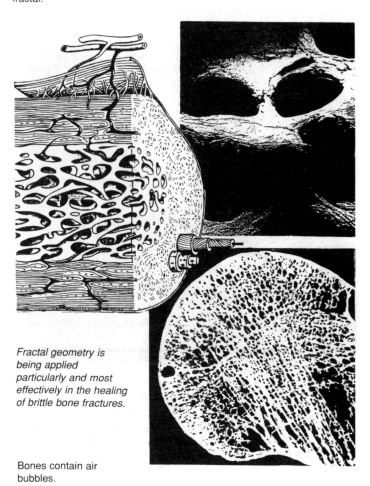

*Fractal geometry is
being applied
particularly and most
effectively in the healing
of brittle bone fractures.*

Bones contain air
bubbles.

Fractal Beats

The body structures of all of nature's animals are fractal, and so too is their behaviour (see Orchid Fractals) and even their **timing**.

Our heart beats seem regular and rhythmic, but when the structure of the timing is examined in fine detail, it is revealed to be very slightly fractal. And this is very important.

Our heart beats are not regular. There is always a tiny variation.

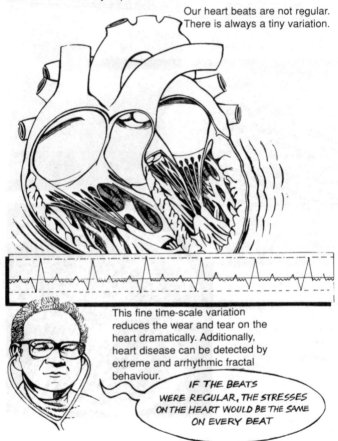

This fine time-scale variation reduces the wear and tear on the heart dramatically. Additionally, heart disease can be detected by extreme and arrhythmic fractal behaviour.

IF THE BEATS WERE REGULAR, THE STRESSES ON THE HEART WOULD BE THE SAME ON EVERY BEAT

Practical Solutions

In the early days, Benoît Mandelbrot found the most enthusiastic responses to his ideas among applied scientists working in practical fields – with oil, rocks or metals. Now all that has changed.

Fractals have become a mainstay in the study of the structures of polymers and ceramic materials, and in tackling the more unsavoury problems of nuclear reactor safety.

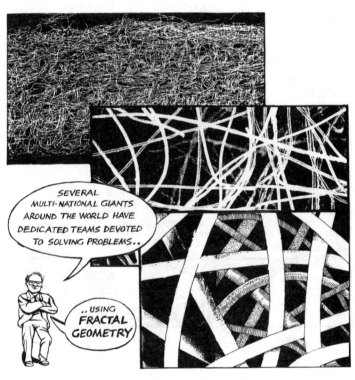

polymer structure in increasing enlargements

Striking Oil

A fresh and novel approach to a variety of geological and geophysical problems using the derivation of the fractal dimension of sand-shale sequences, fault patterns, deltas and channel systems is now in place and proving useful.

The fractal content of 2-D images of fault patterns has resulted in the representation of such images by relatively few equations, which are then iterated. The decomposition of 2-D images to much smaller amounts of data than would be required to represent the pixels, and their subsequent recomposition is vital. The possibility of recomposing the 2-D image at greater resolution leads to the exciting possibility of fault prediction even below the limit of seismic resolution.

Guy Nason and Alan McKeon

FAULT TYPES

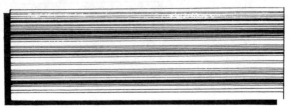

123

The Spring is Sprung

High wastage has always been an intractable and irritating problem for the spring industry throughout the world. 10 per cent of spring wire, which compares favourably with other wires in appearance, with consistent strength and malleability, will not make springs. And the reason for this is the fractal structuring of the molecules within the wire.

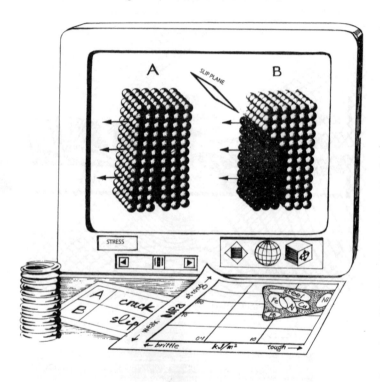

Computer technology can now model the structure of wire and test it, assess it and characterize it.

Testing a coil of wire previously took up to three days; now, using fractal geometry, the time has been reduced to just three minutes! This constitutes enormous savings and greatly increased efficiency to the spring industry.

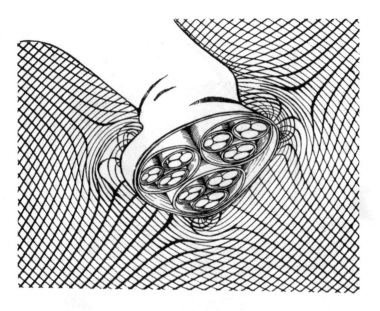

This technology has also found grateful acceptance in processes where strength and malleability are of the utmost importance. It is in use now in the jewellery and fibre-optics industries, where researchers have found fractal fibres, consisting of fibres of fibres, to be the most optically efficient.

Stress Loadings

Statistical models using fractal geometry are in use testing for stress loading on oil rigs and on aircraft in turbulence, with particular emphasis on the effects of very short gusts of wind.

Aerials

Fractal structures have now been found to provide the most effective shapes for mobile phone aerials.

Detection

Fractal geometry has been used effectively by the military for the detection of man-made features in natural environments. This geometry has proved to be remarkably effective in finding and tracking submarines or the wake of ships.

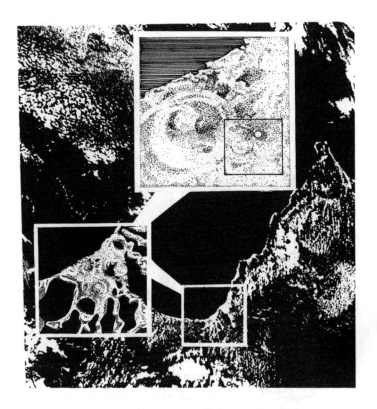

At Queens University in Belfast, **Avian Alexander** has developed a fractal database of shoe-prints which removes the subjective element of identification for forensic scientists.

Fractal Ecology

Fractal geometry is a new mathematical language. We all see fractals every day, everywhere we look. They are already very familiar to us. So it is not surprising that fractal geometry is finding a host of applications in the study and the management of our environment. Acid rain is a prime example.

RAINFALL IS FRACTAL, SO POLLUTANTS CAN BE TRACKED

Earthquakes show up with a clearly fractal signature, as do epidemics in the human species.

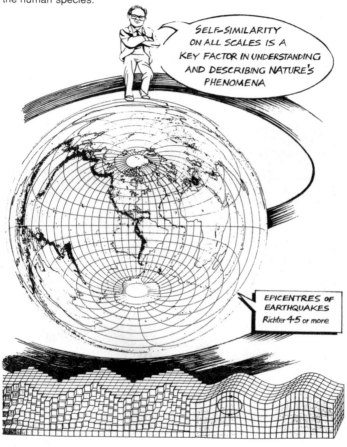

Corrosion reveals the fractal nature of the process, suggesting ways and means to alleviate the problem.

Orchid Fractals and Crowds

Fractal geometry is now used in the modelling of crowd flow. Our behaviour *en masse* has turned out to be fractal. **G. Keith Still**'s work has led to some unexpected revelations about human crowd behaviour and how to manage crowds.

Our behaviour resembles flocks of birds or shoals of fish as they move as one unit.

Crowds entering, leaving and moving around inside stadia reveal remarkable behavioural patterns. Although we are individuals and can make decisions about which direction we should move, how far and how fast, in tight, busy, fluid situations we behave like one organism.

In a crowd our focal point becomes the bubble of our own space and what little we can detect beyond that. Running crowd-flow computer simulations produce extraordinarily beautiful patterns, which because of their fractal nature, have structure on all scales. And, again these patterns appear organic. They are also very beautiful.

Flow simulations very often closely resemble floral patterns like orchids. Still's work is used in stadium design and in the intricate modelling of ecosystems.

Olbers' Paradox

One of the most beautiful fractals in nature is the night sky. Wherever you look, there is a star. Between any two stars there are always other stars. In 1826, **Wilhelm Olbers** argued that in a large enough universe, the sky ought to be uniformly bright.

Since luminosity decreases with the square of the distance, and so does apparent size, the total amount of light coming from any direction in space ought to be the same.

Back in 1909, **Jac Fournier** suggested a model to tackle this problem of the distribution of matter in the universe. He said: take five points, arranged in a square with one in the middle. Replace each point with a scaled-down copy of the whole pattern. Don't stop there! Keep going, replacing each point of this new diagram with a smaller duplicate of the whole shape. And so on . . .

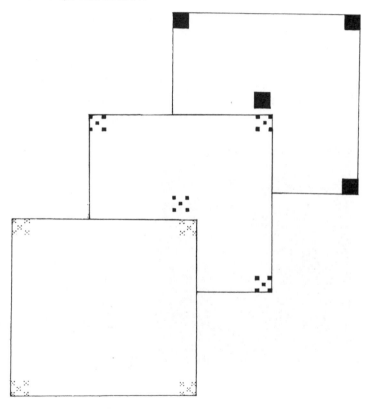

In the limit, you are left with a Fournier fractal. Although highly simplified, it shows that an infinite universe need not be equally bright in all directions, if it has a fractal structure.

The Great Wall

Planets clump together to form solar systems. The stars fit together to form clusters of stars. Clusters of stars gather together to form galaxies. Galaxies hang together to form clusters. These clusters go on to make super-clusters.

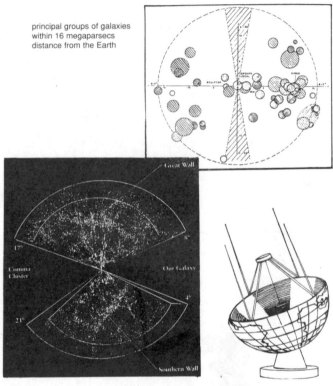

principal groups of galaxies within 16 megaparsecs distance from the Earth

In 1986, astronomers discovered that some galactic clusters are composed of vast agglomerations of matter a billion light years across. In 1989, a huge sheet of galaxies known as The Great Wall was discovered by **John Huchra** and **Margaret Geller** at the Harvard-Smithsonian Center for Astrophysics.

The real geometry of the universe in three dimensions is like a foam. All of the matter in the universe – the stars, the planets and all the other bits (known and, probably, unknown) – is on the bubbles in the foam. In between the galactic clusters there are huge voids with absolutely nothing at all in them.

Using fractals, researchers can now model the evolution and the structure – even the fate – of interstellar dust clouds.

The Big Bang

The universe has turned out to be fractal on scales up to 100 million light years, with a fractal dimension between 1 and 2. Now that we know that it is fractal on all observable scales, cosmologists can reconsider in this new light one of the major problems surrounding the "Big Bang" theory of the universe.

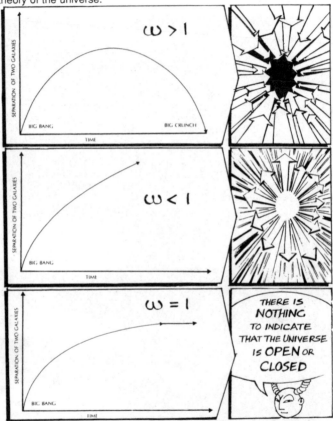

The Greek letter omega represents the ratio between the observed cosmic mass density and the critical density that would cause gravity to collapse the universe back in on itself. It turns out that the universe is critically poised, its omega equal almost exactly to 1. It's "flat".

Many cosmologists and physicists, including Stephen Hawking and his colleagues, have admitted that a universe with an omega equal to 1 poses enormous problems for the Big Bang model.

There are many other issues around the Big Bang theory, which still require resolving. **Hannes Alfvén**, the Swedish Nobel Laureate and originator of the Plasma Theory of cosmology, has put it this way.

Cosmic Connections

Just as fractal geometry is now enabling cosmologists to throw new light on these ticklish problems, the latest versions of the inflationary Big Bang theory imply that the universe is in fact a self-generating fractal, endlessly creating other universes out of itself. This would indicate that it is not just the structure of the universe that is fractal, but its evolution as well.

In this inflationary theory, a chain reaction gives birth to a fractal structure of universes within universes. In this case, there would be no end to evolution on all scales.

How Do We See?

Our brains produce images of far higher resolution than the signals recorded by our eyes. Retinal resolution is limited by the quality of the image provided by the lens and the cornea; by the wave nature of light itself (in the same way that the resolution of a telescope is limited), and by the separation and size of the photoreceptors which pick up the image.

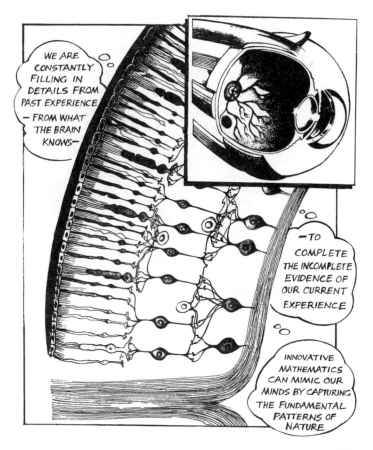

139

Fractal Image Compression

Michael Barnsley believes that our visual cortex uses some kind of fractal algorithm to optimize sensitivity to important features and details in our visual field.

Using a sophisticated computational method, Barnsley has developed a form of image compression which can shrink images to a fraction of their size, and then blow them up to any size without pixellation.

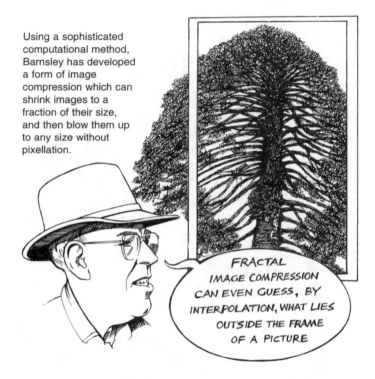

FRACTAL IMAGE COMPRESSION CAN EVEN GUESS, BY INTERPOLATION, WHAT LIES OUTSIDE THE FRAME OF A PICTURE

A team of scientists at the Georgia Institute of Technology, led by Barnsley, discovered a method for reproducing even very complicated forms realistically by a process called "affine transformations", which looks like the original.

Affine Transformations

REAL-WORLD IMAGES HAVE REDUNDANT INFORMATION. TREE BARK HAS REPEATING PATTERNS WITH SMALL VARIATIONS

IF WE COPY AND EXPAND A SECTION..

.. STRETCH IT...

...AND ROTATE IT..

..IT CAN LOOK LIKE OTHER SECTIONS OF THE BARK

The process of rotating, stretching, and moving is called an **affine transformation**.

Any affine transformation can be expressed in mathematical terms with just a few numbers, called an affine map. These numbers are coefficients that are plugged into a standard equation.

141

Morphogenesis

The implications that affine transformations might have for **morphogenesis** (the way form develops in living organisms) in the real world are yet to be explored. But in immediate practical terms, scientists are hoping that these transformations will allow them to derive efficient ways to store complex data in digital memory, transmit photographs over phone lines, and simulate natural scenery by computer.

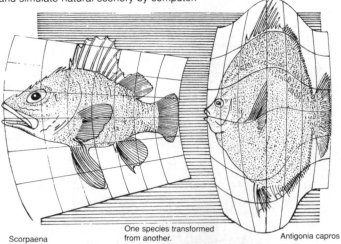

Scorpaena

One species transformed from another.

Antigonia capros

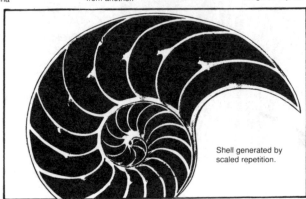

Shell generated by scaled repetition.

Satellites

Weather satellites have been in operation and providing us with useful data for decades. The spy or reconnaissance satellites which patrol the earth produce images of areas of specific military interest with literally thousands of times the definition of the weather satellites! These satellites are required to transmit enormous amounts of data to the ground.

Data compression – the ability to compress images, transmit them, and then expand them again at the receiver – is now a top-priority military requirement.

The Money Markets

Fractals have turned out to be just as ubiquitous in economics as in the natural sciences. Benoît Mandelbrot published his most recent work in 1998. It is called *Fractal and Scaling in Finance: Discontinuity, Concentration, Risk*. It is another dense and impressive book. And, as with his other works, it is very challenging. Built on his previous deep insights, it engages a new world of mathematical problems. It was reviewed by the *Scientific American* under the front page banner headline: A FRACTAL WALK DOWN WALL STREET . . .

This is how Mandelbrot distils the essence of his argument.

The simplest models of price variation assume the most straightforward form of randomness, the randomness of a sub-atomic particle in motion; physicists call this Brownian motion, but in financial markets, when applied to fluctuating prices, it has been called 'a random walk down the Street'. Whether simple or refined, randomness is an intrinsically difficult idea that seems to clash with powerful facts or intuitions. In physics, it clashes with determinism, and in finance it clashes with instances of clear causality – that a price jumped because of some particular event.

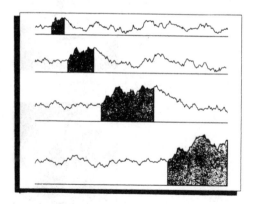

Brownian motion in one dimension at different scalings

Two randomised
Sierpinski gaskets

What Are the Rules?

Mandelbrot had always been convinced that his researches had shown that probability, statistics and fractal geometry could really help describe mathematically what goes on in markets. There were underlying rules to be uncovered and unmasked.

An extensive mathematical basis already exists for fractals and multifractals. Fractal patterns appear not just in the price changes of securities but in the distribution of galaxies throughout the cosmos, in the shape of coastlines and in the decorative designs generated by innumerable computer programs. A fractal is a geometric shape that can be separated into parts, each of which is a reduced-scale version of the whole.

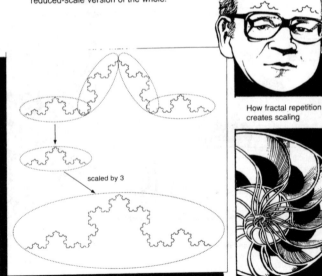

How fractal repetition creates scaling

scaled by 3

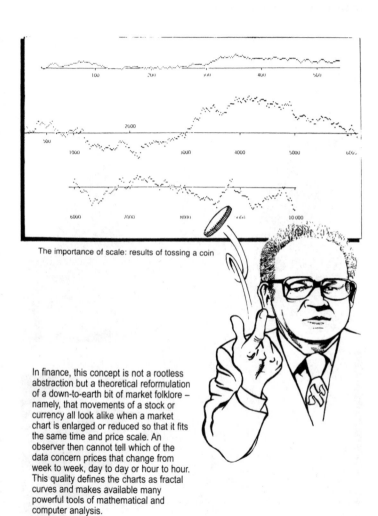

The importance of scale: results of tossing a coin

In finance, this concept is not a rootless abstraction but a theoretical reformulation of a down-to-earth bit of market folklore – namely, that movements of a stock or currency all look alike when a market chart is enlarged or reduced so that it fits the same time and price scale. An observer then cannot tell which of the data concern prices that change from week to week, day to day or hour to hour. This quality defines the charts as fractal curves and makes available many powerful tools of mathematical and computer analysis.

Market Self-Affinity

A more specific technical term now for this resemblance between the parts and the whole is **self-affinity**. This property is related to the most commonly-known facet of fractal geometry, which we have already discussed – self-similarity – in which every feature of an image is shrunk or expanded by exactly the same ratio. Financial market charts, however, are – as we all know only too well – a long, long way from being self-similar.

In a detail of a graphic in which the features are higher than they are wide – as are the individual up-and-down price ticks of a stock – the transformation from the whole to a part must reduce the horizontal axis more than the vertical one. For a price chart, this transformation must shrink the time-scale (the horizontal axis) more than the price scale (the vertical axis). The geometric relation of the whole to its parts is said to be one of self-affinity. The existence of unchanging properties is not given much weight by most statisticians. But they are beloved of physicists and mathematicians like myself, who call them invariances and are happiest with models that present an attractive invariance property. A good idea of what I mean is provided by drawing a simple chart that inserts price changes from time 0 to a later time 1 in successive steps. The intervals themselves are chosen arbitrarily; they may represent a second, an hour, a day or a year.

Generators

The process begins with a price, represented by a straight trend line. Next, a broken line, which Mandelbrot calls a generator, is used to create the pattern, which corresponds to the up-and-down oscillations of a price quoted on the financial markets.

The generator consists of three pieces that are inserted (interpolated) along the straight trend line. (A generator with fewer than three pieces would not simulate a price that can move up and down.) After delineating the initial generator, its three pieces are interpolated by three shorter ones. Repeating these steps reproduces the shape of the generator, or price curve, but at compressed scales.

MOVING A PIECE of the fractal generator to the left ...

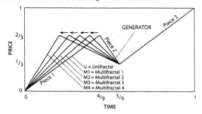

... causes the same amount of market activity in a shorter time interval for the first piece of the generator and the same amount in a longer interval for the second piece ...

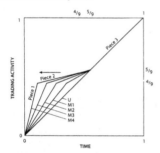

Both the horizontal axis (timescale) and the vertical axis (price scale) are squeezed to fit the horizontal and vertical boundaries of each piece of the generator. Only the first stages are shown in the illustration, although the same process continues. In theory, it has no end, but in practice, it makes no sense to interpolate down to time intervals shorter than those between trading transactions, which may occur in less than a minute. Clearly, each piece ends up with a shape roughly like the whole. That is, scale invariance is present simply because it was built in. The novelty (and surprise) is that these self-affine fractal curves exhibit a wealth of structure – a foundation of both fractal geometry and the theory of chaos.

... Movement of the generator to the left causes market activity to become increasingly volatile.

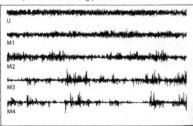

Beyond Portfolio Theory

A few selected generators yield so-called unifractal curves that exhibit the relatively tranquil picture of the market encompassed by modern portfolio theory. But tranquillity prevails only under extraordinarily special conditions that are satisfied only by these special generators. The assumptions behind this oversimplified model are one of the central mistakes of modern portfolio theory. It is much like a theory of sea waves that forbids their swells to exceed six feet. The beauty of fractal geometry is that it makes possible a model general enough to reproduce the patterns that characterize portfolio theory's placid markets as well as the tumultuous trading conditions often encountered.

Multifractals

The just described method of creating a fractal price model can be altered to show how the activity of markets speeds up and slows down – the essence of volatility. This variability is the reason that the prefix "multi-" was added to the word "fractal".

Mandelbrot maintains that probability, statistics and fractal geometry can and will help describe what goes on in the money markets.

The techniques I propose come closer, not necessarily to forecasting a price drop or rise on a specific day or time, but to estimating the probability of what the market might do and getting ready for it. In other words, a light of order in the seemingly impenetrable thicket of the stock market.

Fractals in Art: Mandalas

As Mandelbrot himself has pointed out: "There is something familiar about fractals. I had this experience immediately: that when I first, first saw them, I was the first person to see them! There was absolutely no way anybody could have seen them before. Yet, after a few days, or sometimes a few hours, even a few minutes, they became almost familiar. I was finding features in them, which I had seen somewhere. So where had I seen them? Well, first of all in natural phenomena, but also certainly in art."

Zooming into the boundary of the M-set, we find smaller and smaller island molecules, surrounded by increasingly intricate circular patterns, evocative of Oriental art, particularly in the meditative designs of Buddhism known as mandalas.

In Mahayana Buddhism, the fractal nature of reality is illustrated in the Avatamsaka Sutra by the metaphor of Indra's net, a vast network of precious gems hanging over the palace of the god Indra, so arranged that if you look at one you see all the others reflected in it.

IN EVERY PARTICLE OF DUST THERE ARE PRESENT BUDDHAS WITHOUT NUMBER

We know that different regions in the brain process shape, colour and motion. Benoît Mandelbrot has hypothesized that perhaps there is a specific circuit in the brain to deal with fractal complexity.

Decorative Patterns: Self-Similarity

The abstract, curvilinear motifs of ancient Islamic decorative art found in mosaics and carpet design appear again and again at all scales of magnification on the boundary of the Mandelbrot set.

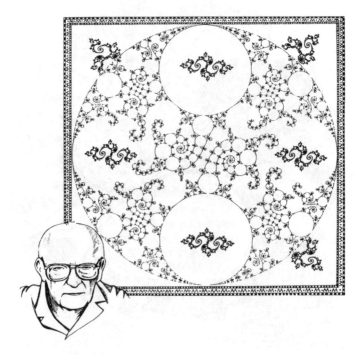

Arthur C. Clarke thinks that it may be just a coincidence, "but", he writes, "the Mandelbrot set does indeed seem to contain an enormous number of mandalas or religious symbols, which are found in ecclesiastical designs – such as stained glass windows, and particularly in Islamic art. We find many forms like the Paisley pattern echoing the Mandelbrot set centuries before it was discovered!"

Fractal shapes were being expressed intuitively by artists long before they were recognized in science. Self-similar patterns appear in Celtic artefacts, like the spirals and circles within circles of the exquisitely crafted illuminated pages of the early 9th-century *Book of Kells* and the Densborough mirror made in the 1st century AD. Mathematical awareness, particularly fractal awareness, reveals itself in the art of the Romans and the Egyptians, and in the work of the Aztec, Inca and Mayan civilizations of Central and South America. Shapes highly reminiscent of the Koch curve were used to depict waves by the Hellenic artist in a frieze in the ancient Greek town of Akrotiri.

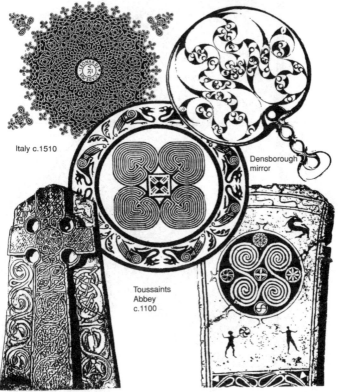

Italy c.1510

Densborough mirror

Toussaints
Abbey
c.1100

Aberfemno, Scotland

Gotland Stone, Sweden, 400 BC

Scaling and Repetition

"The Great Wave" at Kanagawa by the Japanese artist **Katsushika Hokusai** (1760–1849) and **Leonardo da Vinci**'s (1452–1519) "Deluge" portray a profound sensitivity to the dynamics of water, with its ever-diminishing detailed repetition.

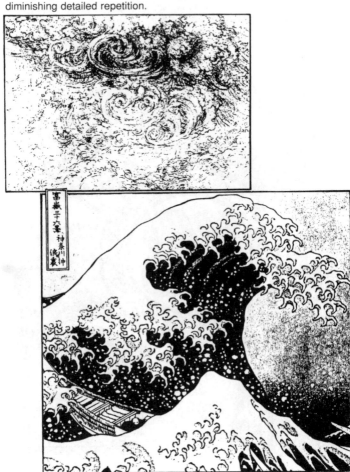

One of mathematician Ian Stewart's favourite works of art is Hokusai's woodcut "Waterfall in Yoshino", in which "the central form is a claw shape, which appears at various scales and in numerous transformations – in the vegetation, in the water, in the rocks, in the horse and in the two straining men. This recurring form provides a sense of unity, diversity and wholeness to the work."

In more recent times, **Salvador Dalí** (1904–89) and **M.C. Escher** (1898–1972) have both exploited the idea of shapes containing copies of themselves. The paintings of the Abstract Expressionist **Jackson Pollock** (1912–56) can be reliably dated just from their fractal dimension.

Cues in the Landscape

Art is about those things which we do not always immediately recognize or understand. The artist helps us to see things more clearly, revealing previously hidden patterns to us. Art is fractal. Art reflects and expresses the fractal nature of our conscious perceptions of the world, as interpreted by our fractal brains.

Bill Hirst, the British scientist and photographer, wrote: "If you remove cues like the horizon from a landscape scene, it becomes difficult to tell if you're looking at pebbles or rocks or hills".

"There was something there I wanted to understand. It wasn't about order and chaos, but something in between. Then it came to me in a flash – it was about fractals!"

Fractals in Music

Spectral analysis of music from classical to nursery rhymes has revealed a remarkable affinity with patterns in nature, in particular a fractal distribution called 1/f noise, which is found in the sound of a waterfall or waves crashing on a beach.

All music from Bach to the Beatles, even birdsong, is characterized by 1/f noise, displaying the same dynamic balance between predictability and surprise, between dull monotony and random discord. Seen in this light, music is essentially a simulation of the harmony in nature.

Dysfunction in Modern Architecture

A MIES VAN DER ROHE BUILDING IS A SCALE-BOUND THROWBACK TO EUCLID..

..WHILE A HIGH PERIOD BEAUX ARTS BUILDING IS RICH IN FRACTAL ASPECTS

It is entirely conceivable that the low level of fractal complexity in modern inner cities is a strong contributing factor to the high incidence of depression reported in these kind of environments.

These modern buildings – office blocks, tower blocks, factory blocks – never actually function in the true sense of the word. They do not work in the role for which they were intended. They become objects of loathing and derision. They get abused. They get uglier and, in a downward spiral, even less "useful".

Organic Architecture

The unfinished cathedral by **Antonio Gaudí** (1852–1926) in Barcelona is a stunning example of fractal architecture. It is organic. It is rich in detail. It is expressive. It is interactive and riveting. Its whirling, twirling curves and detailed branchings cry "fractal"! The building looks almost alive. This organic, fractal quality is, of course, thematic throughout Gaudí's work.

Fractal Traditions

The same can be said of the Paris Opera House (1875), with its intricately detailed baroque carvings.

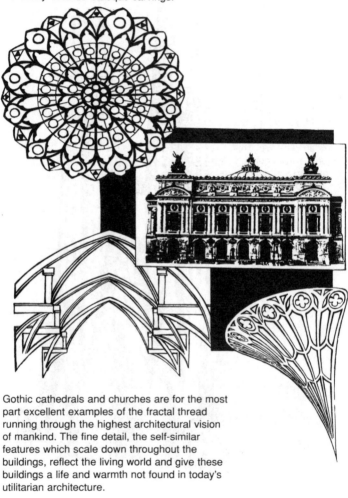

Gothic cathedrals and churches are for the most part excellent examples of the fractal thread running through the highest architectural vision of mankind. The fine detail, the self-similar features which scale down throughout the buildings, reflect the living world and give these buildings a life and warmth not found in today's utilitarian architecture.

Strength in the Details

Architecture is an expression of the human mind as a utility and also as a work of art. Classical architects paid great attention to detail on all scales and to the way their buildings flowed.

There is strength in fractals. A fractal lattice structure provides the maximum strength for any given weight of material.

FRACTAL DETAILS ARE ACTUALLY MORE USEFUL THAN THE FUNCTIONALIST STRAIGHT LINES OF MODERNISM

LOOK AT THE EIFFEL TOWER OF 1889..

Fractal drums make little sound. This is because fractal shapes are the most effective in damping oscillations, suggesting that they might also be more robust than other shapes.

Fracology

Ethnographers have recently uncovered evidence that traditional African societies are modelled, albeit subconsciously, on fractal forms. Aerial photos of traditional African settlements reveal clear fractal structure in the branching of streets, and in the recursive rectangular enclosures and circular dwellings.

Aerial view of Logone-Birni, Cameroon

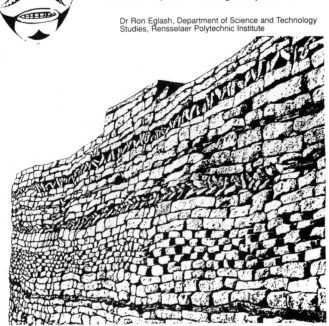

"Recursive scaling structures can also be found in other areas of African culture – in art, religion, indigenous engineering, even in games. In the design rationales and the cultural semantics of many of these geometric forms, there are abstract ideas and formal structures that closely parallel some of the fundamental aspects of fractal geometry. These results agree with recent developments in complex systems theory too, which suggest that pre-modern, non-state societies were neither utterly anarchic, nor frozen in static order, but in reality used an adaptive flexibility, which took full advantage of fractal geometry and of the non-linear aspects of ecological dynamics."

Dr Ron Eglash, Department of Science and Technology Studies, Rensselaer Polytechnic Institute

Quo Vadis?

From 1975, when Mandelbrot coined the word fractal, to 1980, the word appeared in just a handful of academic papers. By 1990, there were 5,000 papers a year published with the word "fractal" in the title.

Here was a new tool that could be used across the board of science, though perhaps, as **Leo Kadanoff** of the University of Chicago has mooted: "Despite the beauty and elegance of the phenomenological observations upon which the field is based, the physics of fractals is, in many ways, a subject waiting to be born. One might hope, and even suspect that eventually a theoretical underpinning will be developed to anchor this subject."

It's early days yet, but asked if fractals are going anywhere, Arthur C. Clarke retorted . . .

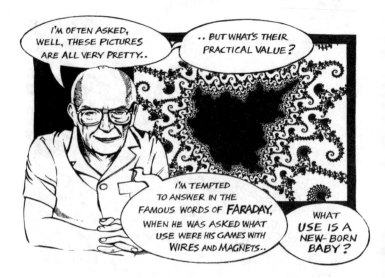

Not Everyone Agrees . . .

In a recent and scathing article in *The Mathematical Intelligencer*, **Steven G. Krantz** of the University of California at Santa Cruz wrote that: "Mandelbrot is good at dreaming up pretty questions. I don't think he's proved any theorems as a result of his investigations, but that is not what he claims to do. By his own telling, he is a philosopher of science."

Mandelbrot had this to say in his defence.

Mandelbrot has always seen himself as a man of action. His geometry is of the real world, of action and reaction, of cause and effect. This is the world of doing – and you can do things with his geometry. "My work and the work of those it has inspired is certainly not found in philosophy journals, but in those of working mathematics, science and art."

The Special Talent of Mandelbrot

Mandelbrot was awarded the Wolfe Prize for mathematics in 1993. The citation reads that the award to Dr Benoît Mandelbrot was for "changing the way we look at the world through the concept of fractal geometry". That is indeed correct. Ian Stewart, mathematician and champion of fractal geometry, has summed up the essence of Mandelbrot's story in his book, *Does God Play Dice*?

The Order Is Out There

We have only begun to see what fractal geometry might achieve for us. This is Ian Stewart's view. "Thanks to the development of new mathematical theories like fractal geometry, nature's more elusive patterns are beginning to reveal their secrets. Already we are seeing a practical impact as well as an intellectual one."

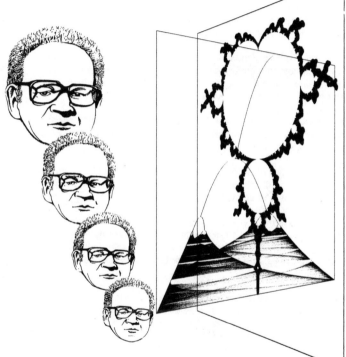

Our new-found understanding of nature's secret regularities is being used to steer artificial satellites to new destinations with far less fuel than anyone had thought possible, to help avoid wear on the wheels of locomotives and other rolling stock, to improve the effectiveness of heart pacemakers, to manage forests and fisheries, even to make more efficient dishwashers. But most important of all, it is giving us a deeper vision of the universe in which we live, and of our place in it.

171

Further Reading

The Fractal Geometry of Nature, by Benoît B. Mandelbrot (W.H. Freeman, New York 1977). This is the seminal work on fractal geometry, and Mandelbrot's first outpouring of his discovery. A remarkable book, but very technical.

The Emperor's New Mind, by Roger Penrose (Vintage, London 1990). Explores the mystery of mind and consciousness and includes a fascinating chapter on fractals and the Mandelbrot set, or – as Penrose calls it – The Land of Tor'Bled-Nam.

Does God Play Dice?, by Ian Stewart (Penguin, London 1990). An entertaining and thoughtful introduction to the new mathematics of chaos theory, which thoroughly explores fractal geometry in this context and traces the historical background to the theory.

The Turbulent Mirror, by John Briggs and F. David Peat (Harper and Row, New York 1989). A highly creative and illustrated guide to chaos and fractals. A lot of fun.

Fractals – Images of Chaos, by Hans Lauwerier (Penguin, London 1991). This slim volume is easy to read, but moderately technical. You need some maths to enjoy it.

Dynamical Systems and Fractals, by Karl-Heinz Becker and Michael Dörfler, translated by Ian Stewart (Cambridge University Press, Cambridge 1989). Covers the mathematics of chaos, fractals and complex dynamics, and is aimed at the technically-minded who are familiar with computers.

The Ghost From the Grand Banks, by Arthur C. Clarke (Victor Gollancz, London 1993). In this imaginative novel about raising a sunken ship from the ocean bed, the author cleverly weaves in a pocket-sized description of the Mandelbrot set.

Leadership and the New Science, by Margaret J. Wheatley (Berrett-Koehler, San Francisco 1992). This book shows how the new mathematics is providing insight into the way we organize the workplace, and how things might be improved in the future. Fractals play a big part in the author's argument.

Fractals Everywhere, by Michael Barnsley (Academic Press, San Diego 1988). One of the first books on fractal geometry. This is a ground-breaking work, but definitely for the mathematicians among us.

Fractals – the Patterns of Chaos, by John Briggs (Thames and Hudson, London 1992). This is a beautifully illustrated coffee-table book which captures and explores the fractal characteristics of the natural world in the light of chaos theory.

The Web of Life, by Fritjof Capra (HarperCollins, London 1996). The author presents a radical synthesis of recent scientific breakthroughs in complexity theory, chaos theory and fractal geometry.

From Here to Infinity, by Ian Stewart (Oxford University Press, Oxford 1987). Described as the perfect guide to today's mathematics, this book is very accessible and is a fun read.

Fractal Geometry, Mathematical Foundations and Applications, by Kenneth Falconer (John Wiley, New York 1990). An in-depth account of the theory of fractals and Hausdorff dimension, and their physical applications.

Mazes for the Mind, by Clifford A. Pickover (St Martin's Press, New York 1992). A visual exploration of the frontiers of computer research into the strange world of Cantor cheese and fractal ant farms.

The Science of Fractal Images, edited by Heinz-Otto Peitgen and Dietmar Saupe (Springer-Verlag, Berlin 1987). A beautiful book containing recipes for many stunning fractal constructions.

Acknowledgements

The authors are infinitely grateful to Ian Stewart and Michael Barnsley in particular, and also to Arthur C. Clarke and Roger Penrose for their guidance, encouragement and support in creating this book. Though our biggest thank you, of course, goes to Benoît Mandelbrot for his courage, his persistence, his good-will and his brilliant discovery.

Biographies

Nigel Lesmoir-Gordon formed his first production company, Green Back Films, in 1976, and worked for Donovan, Pink Floyd, 10cc, Squeeze, Rainbow, Joe Cocker, Big Country and Wings. He later joined the creative team at the Central Office of Information, writing and directing for the international TV documentary series *This Week in Britain* and *Living Tomorrow*. His work includes an acclaimed series of films for the Royal Air Force, *Saving the Children*, a television documentary on women who work for children's charities, *The Bobby Charlton Story* and the series *Whatever You Want* for Channel Four. In 1995 he made the award-winning television documentary *The Colours of Infinity*, presented by Sir Arthur C. Clarke, on the discovery of the Mandelbrot set and the development of fractal geometry. It has thus far been broadcast in over a dozen territories worldwide. He has just completed *Is God A Number?*, a television documentary looking at the mystery of consciousness and some remarkable discoveries that have recently been made in mathematics.

Will Rood was awarded Senior Optime in mathematics by the University of Cambridge, and received his MA there in 1992 for work on transfinite set theory. The previous year he set up the design and software company, SoundNatureVision, producing powerful music and graphics packages for the RISC-OS platform. Programming in assembly language, he began to explore the strange and beautiful world of fractals and cellular automata brought to life by his versatile visualization application, !NatureVision. His fractal animations have graced many television documentaries, featuring extensively in Nigel Lesmoir-Gordon's *The Colours of Infinity* and *Is God a Number?* and Channel 4's *Equinox*, as well as in pop videos by artists as diverse as The Infinity Project, Mike Scott and Star Sounds Orchestra. His artwork adorns numerous magazines, CD sleeves, posters and T-shirts. *Infinit*, a visual exploration of fractal geometry, was released on video in 1997. Current posts include Science Editor at *Dream Creation* magazine, and consultant on Ron Fricke's Imax project, *The Infinite Journey*.

Ralph Edney is the author of two graphic novels, and illustrator of *Philosophy for Beginners*. He is also a cartoonist, illustrator and cricket fanatic.

Index